Good Wood

Good Wood

Growth, Loss, and Renewal

Steven R. Radosevich

Oregon State University Press
Corvallis

The paper in this book meets the guidelines for permanence and durability of the Committee on Production Guidelines for Book Longevity of the Council on Library Resources and the minimum requirements of the American National Standard for Permanence of Paper for Printed Library Materials Z39.48-1984.

Library of Congress Cataloging-in-Publication Data
Radosevich, Steven R.
Good wood : growth, loss, and renewal / Steven R. Radosevich.
p. cm.
ISBN-10: 0-87071-115-6 (alk. paper)
ISBN-13: 978-0-87071-115-2 (alk. paper)
1. Farm life—Washington (State)—Yakima River Valley. 2. Farm life—Oregon—Willamette River Valley. 3. Radosevich, Steven R. 4. Yakima River Valley (Wash.)—Social life and customs. 5. Willamette River Valley (Or.)—Social life and customs. I. Title.
S521.5.W2R33 2005
630'.9797'55—dc22
2005008737

 First edition 2005
Printed in the United States of America

Oregon State University Press
500 Kerr Administration
Corvallis OR 97331-2122
541-737-3166 • fax 541-737-3170
http://oregonstate.edu/dept/press

This book is dedicated to

the Radosevich and Clemans families, my Good Wood

Contents

Preface

Every Sunday morning a group of farmers gathers at Newland's drug store in Tieton, Washington. These old men have been doing this for as long as I can remember; since I was born, maybe longer. They pick up the Sunday newspaper, drink coffee at a back corner table, and talk about the winter's freeze, Jake's new tractor, or where their kids and grandkids are. Bumper crops, springtime freezes, their kids' growing then moving away, and this town they call home link them all. Each of us has such a place that we belong to. It might be the house you grew up in, a trail along the river, or a grove of trees no one else knows about. It could be a place you always find obsidian, buy string, or catch crawdads. It's a place where people you love grow old, where places that used to be have changed but you'd like to see again.

If you ask where I'm from, I'll tell you about the Pacific Northwest; its evergreen forests, broad rivers that flow through fertile irrigated valleys, and a great high desert covered with cheatgrass and sagebrush. It was once a place of one-log loads, company towns, and an ancient forest so vast I thought it could never be taken away. It's where the Columbia River flowed free over the falls at Celilo, and salmon were as abundant as the trees of the forest.

I am a scientist, an ecologist, and a professor of forestry at Oregon State University. I teach about forests and farms and a philosophy of natural resource use. I am also a farmer, and this book is about choices—some measured and intentional, while others are not. I'll tell you about my family's farm, the homeplace, at the head of the Yakima Valley. Grandpa bought the place from its homesteader, plowed out the sagebrush, and planted it to orchards nearly a century ago. I've touched the dry graves of forgotten settlers who lived in the desert above this farm. I learned to prune trees in my grandfather's orchard where every cut was a decision, conscious and deliberate, about the health of the trees and their coming crops. I felt the farm's pulse when Grandpa and Dad farmed together, then saw it wither and nearly fall apart. I watched Dad nearly lose the homeplace to winter freezes, low prices, and debt. I saw him almost not care and nearly accepted losing it myself as just the way of farming.

The ancient forest has been cut now, and I've planted firs and pines back in the tracks of bulldozers. But clearcuts don't make forests; they only replace them. There's too much good wood cut away, lost, for it to be otherwise. I watched Celilo Falls sink into the Columbia to be replaced by closed spillways and cheap aluminum foil. I've counted what's left of the salmon through a Plexiglas window inside Bonneville as each fish climbed over the dam on a stairway toward extinction. I've seen farmers caught on agriculture's treadmill, where every new technology costs more and grows more food, but lowers market prices. Everyone benefits from the treadmill except those on it, so there are fewer old men to hang around places like Newland's drug store anymore. All these losses have been decisions too, made consciously I think, but with less deliberation than the forethought of a farmer pruning his orchard.

Salmon can swim free once more if we care enough. Forests will grow again if we have enough patience, humility. Farms are still tended by those who love the land. I've traveled to many places but always return to Oregon. Lynn, my wife, and I call our vineyard here Kla-kla-nee—the Indian name for the three Cascade volcanoes that we see across the Willamette Valley from our home. We planted the vineyard ourselves, and we spend most weekends of our rainy winters pruning it. Every cut is a loss; a lot of wood, good wood, gets laid on the ground. But each cut is also a choice that improves the vitality of the vineyard and growth of our new crop. Our kids, now grown, return every year to help us prune and harvest. And each spring, we gather with my brother and sisters in Tieton to work on the homeplace once again. We say the trees we've planted there are an investment in the children of our grandchildren.

These are my stories, essays about the places I call home.

SRR
Corvallis, Oregon
spring 2005

Acknowledgments

I started this project nearly ten years ago with no idea of writing a book. Back then, I only wanted to write better and be more creative than the scientific journals that I wrote for allowed. I also hoped to put my observations and reflections about farming and forestry to paper for myself as much as anyone. Lynn, my soul mate and wife of thirty-seven years, provided perennial support during this entire time. She's lived the experiences I write about as much as I have. Thank you, Lynn.

Our children, Steve, Susan, and Matthew, daughter-in-law, Teresa, son-in-law, Steve, and grandsons, Tyler, Tim, Kyle, and Nick, and any grandchildren yet unborn provided the inspiration to write these essays. These are the stories I want you to know.

Brother, Joe, and sisters, Babi, Carol, Annie, and Polly, you are much of the good wood in this book. Without your help, many of these stories would not have happened. *Good Wood* is our continuing story. Josip and Paulina, George and Nina, Dad and Mom, thank you for leaving us the homeplace; it's the place we come back to.

I thank Chris Anderson for introducing me to the essay as a form of expression, and Lex Runciman for always searching for more detail, more detail. I also must graciously thank my current writing group, including Charles Goodrich and Gail Wells, for their patience, advice, and encouragement on the many drafts of this book. I still continue, after a decade, to look forward to our meetings and to learn from each of you. Thank you, Mary Braun,

OSU Press, for believing there was a book hidden amidst all these stories.

The Pacific Northwest has been embroiled in an environmental debate for thirty years over how to use and sustain our natural resources. I have been in that debate every day of my professional life and have seen it from many sides, as scientist, farmer, forester, ecologist, and teacher. Lynn says in one of the essays that I like to work alone in my vineyard. True. It gives me a chance to sort out my feelings about this land and our debate over how it is used. Some of you reading these pages may see yourselves there. If so, my words are written with hope, reconstruction, and reconciliation in mind. It is time for resolution to begin.

Part One

Growth

I grew up in a place called Tieton (pronounced *Tie-a-ton*). It's both a town and an area. The town was founded in the early 1850s when the residents there got a post office located in their new two-story grocery store and dancehall, next to the blacksmith's shop. When I was a kid in the 1950s and '60s Tieton had a theater, two barbershops, three taverns, a cafe, drugstore, bowling alley, and two grocery stores—one with a meat locker. It also had a hardware store, two churches, and a feed store, all surrounding a one-block-square park with a gigantic fir growing in the middle that is still the town's Christmas tree. These days the buildings are still there but there's not much in them, except for the post office.

The Tieton is a high basalt plateau in the upper part of the Yakima Valley. It's wedged among the foothills on the east side of Mt. Rainier and Mt. Adams. Cowiche Creek and the Tieton River dissect it east to west. Windblown soil made from volcanic ash covers the plateau and sagebrush and cheatgrass-covered outcrops of basalt boulders stud its landscape. The native sage and bunchgrass between these "rock piles" was cleared away a century ago. The area is orchard country now, mostly apples.

Our farm, the homeplace, is a mile and a half out of town. It surrounds one of the basalt buttes on two sides and borders the North Fork of Cowiche Creek on another.

Homeplace

It's early spring and late in the day. The drowning sun at our backs casts half-light onto the basalt bluffs of the Columbia Gorge as Lynn and I turn east on to Interstate 84 outside Portland. The Columbia is seldom calm, its narrow cliffs often pulling foul weather upstream, churning the river into unruly whitecaps for over a hundred miles. But this particular evening its winds lie inexplicably dormant and the failing light makes the cheatgrass hills rolling ahead of us into Washington's cold desert seem greener, more fresh, somehow. We are on our way to the homeplace, Dad's farm near Tieton in the Yakima Valley. It's been my apprehension, but also resolve, about the homeplace that's troubled me for so long. I'll take over my father's farm tomorrow, but I feel strangely at ease, like the quiescent winds of the Gorge—at truce with myself temporarily about the old farm and what it became. Grandpa, Dad's father, bought the place from its homesteader. He and Grandma grew hay and apples and raised five kids there. Dad, their youngest, took it over when Grandpa died. I hope I can do as well in keeping the link between my family and our land.

We pass Bonneville Dam, the place Lewis and Clark almost two hundred years ago called the Great American Chute. Half a mountain slipped into the Columbia here, making it so narrow that only the most reverent dared challenge the river's swiftness. Indian legend says there was a land bridge across the river at this place, a bridge of the gods. Here, the mighty Columbia cuts through the Cascades and here the Corps of Engineers finally did what even the forming mountains could never do, stop the river. Upriver, we pass Celilo Falls, where now only a road sign points to this ancient Indian fishing ground, submerged by The Dalles Dam. The Indians built wooden platforms that extended out over the river's rocky edge. They fished for salmon in the rapids with nets hung on long poles. When I was little, Granddad, my mother's father, bought salmon out there on the rocks from the Wasco and Warm Springs tribesmen. It was a dusty full day's drive for him to Celilo from his farm near Tieton. He paid fifty cents for a washtub full. Grandma canned some, then gave the rest of the fish away to friends and family. Now the Falls are covered over and the culture built around them is inundated too, submerged and nearly invisible.

We turn north, cross the Columbia on a steel bridge, and climb a crooked grade into the high pines and sagebrush scrub of the Satus. Here, Yakama horse soldiers and U.S. cavalry, five generations ago, some led by Lynn's great-grandfather, fought over this land. Now fencerows, gravel roads, and hayfields break up the native prairie of bunchgrass and sage. We pass an abandoned farmstead along the way. Its worn-down buildings, now covered with wild rose and surrounded by black locust trees planted a century ago, are a legacy in the West; a reminder of our continuing story of hope, use, and sometimes abuse of this land and its people.

The sun sets and we drop into the desert night. A sudden thunderstorm splatters rain against our windshield, but the shower is short-lasted. "Hardly enough to settle the dust," Dad would say; his gravelly voice breaks into my thoughts for a moment. The rain makes the air smell fresh, full of ozone and the aroma of new-growing sage. I crack a window and more scent from the desert rushes in. We breathe deep, holding the smell of home in our lungs for as long as we can. Later, Lynn, knowing my apprehension, says, "Steve, you're a farmer in the heart. You'll know what to do with your Dad's place." I nod, but am still uncertain about what the next day will bring and what I will do after that. We continue on silently, following our headlights and the moonlit outline of Ahtanum Ridge into the Valley.

Over three decades ago I stumbled across another old homestead like the one Lynn and I passed as we crossed over the Satus. It happened in the Wenas, the dryland hills above the homeplace, while I was hunting doves with my high school friend Don. When all the good river and creek bottoms had been taken in the Yakima Valley, settlers staked out homesteads near ephemeral springs in the low sage-covered foothills of the Cascades; places like Tieton, the Satus, and Wenas. The seeps flowed year round during good years but usually dried up during droughts. It was the only land left in the desert that had any water on it.

It was August and during the last heat of the day when Don and I, surrounded by a plain of dry wheat stubble and basalt outcrops, set out cross-country along a barren creek bed we'd found several weeks earlier. After a half-hour's hike into the drylands, we located a patch of scraggy cottonwoods in the distance. It would be easy shooting, we figured, if we could get there before dark—before the doves returned to roost after feeding all day in the grain fields. Never straying much from the washout, we wound our way between the stubble-covered hillsides for another half-hour. Then we came to it, a homestead concealed beneath the rolling hills and marked only by the cottonwoods we'd been hiking toward. It was abandoned and apparently forgotten except by the wheat farmer who'd carefully plowed around the forlorn homesite.

There were two small buildings constructed of rough-sawn boards, their roofs collapsed, caved-in. The walls were still upright but leaning inward. Who built these houses in this desolate place, I wondered? We peered inside the broken-down door of one of the shacks. Nothing was there but the remains of a rough stone fireplace of black basalt. The sagebrush had already reclaimed its earthen floor. We found the slumped-in sides of a hand-dug well near the old cottonwoods. It was deep enough to fall into, but disguised now by a layer of rattlesnake weed and cheatgrass. We sat cross-legged under the sagebrush next to the old well and waited, shotguns cradled across our laps, for the doves to fly in. It was dusk when through the low sagebrush stems and failing light, we saw, unmistakably, the graves—hidden, themselves, beneath the sage. Unsure of our discovery, we moved closer, staring in the failing light to get a better look. There were three of them, side by side, but no markers. Two were little and one was bigger; all were grown over, hidden beneath a quilt of dried-up cheatgrass stems and fallen sagebrush leaves. "Are these real?" Don half-whispered as if someone else might be listening in. Kneeling to get closer, I touched the middle grave with the palms of both hands. I felt the rough basalt stones arranged evenly over the low mound, the rocks themselves now buried beneath a layer of wind-blown soil and dried grass. Who lived here, I wanted to know, and how did they die? If they could tell their story, would it be of hope or only desperation and despair?

That evening while sitting beside the three lost graves in the Wenas, I knew that such places are sacred—the cottonwoods, creek, graves, this land—all sacred. Made sacred by the people who lived there, by the hopes begun and ended there, and by the sacrifices made, buried, and finally forgotten there. What more can be done than die there to finally call such places ours?

I awake early the next morning and drive alone from Lynn's parents' home in Yakima to the homeplace, intentionally taking the long way so I can study the farm from atop a small rise on the highway before I actually get there. The rock butte, a dry hill of basalt boulders, gray sagebrush, and last summer's cheatgrass, hasn't changed. Wild violets, sage buttercups, and a new crop of cheat are up there, I know, hidden in last year's brown duff. Groundhogs, still in winter's slumber, wait for milder weather in tunnels carved among the boulders. The roof of Dad's house stands out at the base of the butte. A single maple that Grandpa planted over eighty years ago, now with limbs as big as barrels but naked from the winter, hangs over the old house. A willow-choked creek wanders across the farm's southern edge, then turns east to where Orie Comet's ramshackle farm used to be. I've fished for rainbows in that creek since I was six years old, sometimes by myself but more often with Dad or Uncle Al. I taught my kids how to fish there. Forty acres of apple, pear, cherry, and apricot orchards once followed the farm's contours between the butte and creek. The orchards should be blooming by now with acres of pink or white blossoms covered by the buzz of wings, bees gathering nectar and pollen for their hives. But only the last few acres of cherries, the trees hardly alive, remain.

Dad sits in his shirtsleeves, waiting, drinking coffee at the kitchen table as I come in the backdoor without knocking. He's lived alone ever since he and Mom divorced, nearly twenty years earlier. He's grown a beard since I'd seen him last and someone, probably one of my sisters, had trimmed his thinning white hair. He looks up and smiles a greeting but doesn't speak. It isn't going to be easy for him to sign the quitclaim today that turns over the homeplace to me, but it has to be done. His last orchard is dying. He can't water it any more; the old irrigation system is rotting away. He's worn down too, tired, past the point of no concern, and being pressured to sell to pay back taxes. The county and irrigation district have threatened an auction. He doesn't especially want to sell, I know, only to be out from under the burden the farm has become.

Farming must have been different for Grandpa than it was for Dad. For Grandpa, most of the stuff he needed to live was grown on the farm, used there, and the rest sold. It was a kind of synergy between himself and his land. Grandpa and Grandma grew a garden, raised chickens and hogs, milked two cows each day, and worked the orchards with a team of horses. Later, they bought a gasoline tractor to help out. He and his sons irrigated the orchards and dragged long handlines filled with arsenic spray around each tree. The spray protected the apples from codling moths, the maggots that ruin the fruit. The land and trees responded to the work, creating a cycle—a currency—that supported all of them. But the cycle must have broken by the time Dad took over; farming now is more like a pipeline where chemicals, energy, and money pass through farms from suppliers to consumers. Higher production—more than Grandpa ever imagined—and an urban workforce are the gain; ever lower market prices, fewer farms, and a displaced way of life are the loss. Today, fewer than two percent of us are farmers, down from over eighty percent when Grandpa bought this place. Maybe that's how Dad almost lost the farm; he became a statistic of progress, a misfortune of change, the product of a century-old policy to move farmers off their land. Then again, maybe it was just freezes and bad luck that broke him down.

We drive the mile and a half to Tieton together. The city hall is a squat square building of gray cinderblock that doubles as the town's jail. Dad delays in the Justice of the Peace's office. He wants me to explain again how the legal transfer will happen. I describe the deal one more time. The six of us kids have formed a joint venture, a kind of partnership, to run the farm. He can live there as long as he wants to. None of us will sell the homeplace, never, and we will all share whatever the farm brings in—if it ever does again. Still dragging his feet, he leans heavily against the wooden countertop that separates us from the rest of the office. He looks at me, the Justice of the Peace, then stares across the room. He dawdles. He stalls. No wonder—he's lived on the homeplace his whole life. Grandpa and Grandma moved there when he was seven years old. It's got to be hard: to sign over a lifetime of work, worries, hopes, expectations—even to your children. I hope this transaction represents a resolution for him, another step in his life and transition for the homeplace.

Grandpa and Grandma had a daughter and four sons and all of them, except Dad, eventually left to farm their own places. I remember finding an old King Edward cigar box several years earlier. Someone had carefully tucked it away between the ceiling joists in the cellar of the old house, my grandparents' home, the same house Dad lives in now. The box was filled with yellowed canceled checks, all from Seattle First National Bank and dated October 1945. Grandpa had written each check in pencil to one of his grown sons—apparently their wages—and each check was endorsed, like an autograph from each uncle and my father; a reminder of what the homeplace is to us.

Finally, as if annoyed by an unnecessary delay, Dad snatches the pen from the countertop and quickly signs the paper. He straightens to full height after he scratches out his signature, sighs deeply, and looks straight at me. Deal done! Then he holds out a hand for me to shake, but the gesture turns into a bear hug. He's strong and pulls me to his chest but he seems thinner, less substantial, than I'd remembered. I'd expected more bulk inside his winter mackinaw. Then we release and I sign the paper too.

The Land They Belong To

The floors in this musty old house slant. Its doorways are lopsided. After a hundred years, its foundation of basalt boulders slumps on one side. I sometimes come here to Grandpa's and Grandma's house, sit there in the living room and rock in my Grandma's wooden mohair chair by the window, listening to the chair creak and to the house groan as its tired foundation settles in a little more. I should know more about how both sets of my grandparents lived, who they were before I was born. I should know if they rose early and drank coffee each morning, what color their dishes were, if they read before going to bed. I should know more than just the names of their parents and how many kids they had and when they each died. I should know why both pairs of grandparents settled in Tieton, an isolated little town at the upper end of the Yakima Valley, why they built farms and grew families there, but I don't.

What I do know about the histories of my grandparents comes from brief conversations with my aunt and uncles, and Mom's long-time interest in genealogy to trace down both sides of my family tree. This is the story of two farms and the people who settled on them, where they came from, who their parents were. It's hard for me to separate my grandparents from where they lived, the orchards they planted. The land they lived on is who they were, what they believed in, and what I know most about them. They are who I became—at least partially formed, like the farms they settled, by their genes, their decisions, opportunities, and accidents, and the will to believe in what might be. Here are the stories I know.

Radosevich Homeplace (1923–Present)

This old house, where I sit and rock in Grandma's chair, is where my paternal grandparents, Joe and Paulina, lived for over fifty years, but now its whitewashed plaster walls and linoleum floor are cracked and worn. This house was Dad's home too after he and Mom split up, but it's been vacant for almost a decade, ever since Dad died in 1997. My brother, Joe, and I boarded up all the windows and turned off the well and lights after Dad got sick, when it was clear he couldn't leave the hospital.

What was it like here before I was born, when Dad and my uncles were young? I have a few photos of them, timeworn, some with dates and names scribbled on the back. I found the pictures in the attic with a dusty envelope full of yellowed papers: a marriage license, a bill of sale for ten acres they bought during the Depression, Grandpa's naturalization papers. And I have bits of their stories, conversations overheard, reconstructed. I spoke Croatian when I was three, but I can't now. It was the language of this house until they realized I knew it too. Then they stopped and only talked in English, especially when I was around. They wanted me to be American.

Grandpa, Josip Radosevic, and his older brother, Vinco, left the small farming village of Mrkopalj, Croatia, when they were still young men—around twenty-five years old—well before World War I, when Croatia was still part of Austria. I'm not sure why they emigrated; no one ever told me, but they left in a hurry. Maybe there was a war brewing in the Balkans. Maybe they were about to be drafted; perhaps they were in trouble. I don't know, but they left carrying little and leaving a lot.

The brothers sailed for New York in 1900 and were processed through Ellis Island as many other European immigrants to America were. But the two brothers left New York City soon after they arrived. They added an "h" to the end of their last name and started working their way across the country, stopping only long enough at each new city to earn money for food and train fare to the next western town. First Chicago, so the story goes, then the steel mills of Gary, Indiana, where they stayed for about a year. They finally ended up in the coal fields and forests around Roslyn, Washington. Maybe they decided to stay in Roslyn because the countryside reminded them of the pine-covered hills of their native homeland. Perhaps they stayed because there weren't many other places west to go. They found other Croats living there and settled in, becoming coal miners in Washington's latest boomtown.

Grandma, Paulina Starosevic, also grew up under the same chalk cliffs of Mrkopalj, Crotia. I've seen photos of the town, downloaded from a traveler's website, still tiny with white church steeples nestled beneath its limestone cliffs. I suppose she knew my Grandpa when they both lived there, before he and his brother took off for America, but probably not well; she was eleven years younger than he was.

When Paulina was sixteen, at the bare turn of the last century, she received an exceptional offer—an invitation from a family friend living in New York City, a boat ticket to America, and a marriage proposal. She didn't know the man, at least not well. He was her father's friend. Nevertheless, later that year, Paulina and her sister, Stella, who was a year younger, boarded a ship bound for New York. The girls arrived at Ellis Island in 1902 after about a week's voyage and found jobs working in a laundry in New York's Croat ghetto. But soon Paulina realized that she couldn't marry her intended. She didn't like him! Was he too old or boring? Was he pitiful or just not the right guy? I don't know.

What conversations did the two sisters have, I wonder? What fates did they conjure up for themselves in their lonely boarding house room each night? Did they write home, seek advice, cry? In 1903 engagements were more like legal arrangements than betrothals, and Paulina knew she couldn't pay back her boat fare to get out of the engagement. The girls had only a little money left over from their voyage to America. They'd saved some by working at the laundry, but it wasn't nearly enough.

Even if the girls could have found a way back to the old country, I'm not sure they would have gone. If the notion was there, it must not have lasted long because they pooled what money they had and secretly bought train tickets to a place as far from New York as they could imagine—a boomtown in the coal fields of Washington, three thousand miles away. Had they heard of other Croats living there, in a place that looked like the hills around Mrkopalj? Did they know there would be prairies, mountain ranges, and a desert to cross? Were they worried about strangers, anxious about the trip, or did they look forward to each new town they'd travel through? The girls could barely speak English. They'd be travelling alone across a continent to a place they had only heard about.

Less than a year after Grandma and Stella stepped off the train at the Roslyn station, she married Grandpa. I know because their marriage license is dated January 4, 1904. She was seventeen, he twenty-eight. A few weeks later, Stella married Rudy Krulich, another miner and Josip's best friend.

Josip and Paulina worked hard. She cooked and cleaned for the miners living in a rugged three-story boarding house, while he worked the mines. They eventually bought a house on one of Rosyln's side streets. I saw it once in the late 1950s on a trip to Seattle with Dad and Mom. It was small, but with two stories, and painted white with brown trim. They furnished it sparingly from the mine's company store. I know this because I have their bedroom chest-of-drawers, a simply carved oak stand and mirror dated 1905, Rosyln Company Store, in chalk on its back. Their first child, Steph, a boy named after Grandma's father, died at birth. Then Anne was born, followed by Jack, Al, Tony, and finally Frank—my dad. I have one old photo of them. Grandpa is short, dark, and muscular with salt-and-pepper hair, a stout neck, and full black mustache. He stands stiffly in a dark suit. Grandma sits in front of him, pregnant with my dad. She's small and pretty with fine features and gray eyes. Her coal-black hair is parted down the middle, apparently tied tight in the back. She's dressed in a simple long white dress and holds her latest babe on her lap, Tony, who's dressed in a pure white baptismal gown. Standing next to Grandma is Anne, a slender girl about ten years old who's also in a long white dress. She's holding a ring of flowers. Next to her are my two other uncles, Jack and Al, boys dressed in new black shorts, suit coats, and bow ties. Al's bigger, though he's two years younger than his brother.

Nobody ever talked about the cave-in at the Roslyn mine when I was growing up. I've only pieced this part of their story together from bits of conversations among my aunt and uncles. I learned that Grandpa, then about forty-five years old, was near the bottom of the mine. He'd been on a ledge and lying on his back, chipping coal off the side of a narrow seam when the whole wall collapsed. Did Grandma run to the mine when she heard its sirens blast out the cave-in alarm? Did she wait with other women at the mine's open mouth while it belched out foul smoke and coal dust? Did she hold her children near, wrap them in her skirts as they watched the miners carry up the injured one by one? Or did she wait at home alone, fretting for news that seemed to never come?

Grandpa was among the last of the miners brought up from the cave-in that day. His forehead was crushed and one eye lay out of its socket, his cheekbone fractured. But he was alive. A frontier doctor put his eye back in and then reconstructed his forehead, reforming the torn muscle and skin around a silver plate that he implanted in the head. The surgery was done so skillfully that Grandpa could still read, and no scar ever revealed the depth of the injury. My aunt told me years later that he tried to go back down after the wounds healed, but the scar of the cave-in apparently went too deep. He hated that pitch-black hole, she said. He dreamed of its dank smell and belching coal dust. He remembered the earth rumbling, pain all around, his own head split open, and the thought of being buried there still alive. He never went into the shaft again, nor did he or Grandma talk about it. I learned all this the day after Grandpa died, when the undertaker asked my aunt how Grandpa came by the silver plate planted in his forehead.

Without a job and nearly broke from the accident, Grandpa and Grandma borrowed four hundred and twenty dollars from his brother, Vinco, which they combined with five hundred dollars of their own to buy a farm near Yakima. There were rumors that the government was bringing irrigation water to the desert there. They bought thirty acres about a mile and a half outside of Tieton, a full hundred miles from Roslyn, but less than forty miles by the crow. The farm lies at the base of a sagebrush-covered basalt butte. A creek runs through part of it. It had a well-constructed barn and a two-story house that is no bigger than a good-sized cabin nowadays.

Ten years before Grandpa and Grandma bought the place, the original cabin on the farm burned down. But the neighboring families had a house raising for the homesteader, and he later sold the land and new house to my grandparents. The neighbors built the new house on a foundation of basalt boulders that they pulled from the side of the butte with horses. The house has a cellar, two bedrooms, living room, and kitchen downstairs, and an attic upstairs where Dad and my three uncles slept. Dad was seven years old when the family moved in.

Grandpa and Grandma bought two horses to work the fields and a cow for milk. Grandma made cheese and butter and picked eggs from under her hens each day. Grandpa planted orchards of apple and pear in the deep volcanic soil and watered the trees from wooden barrels hauled up from the creek until the irrigation came in. It was 1923.

Grandpa and Grandma's marriage lasted over fifty years, until 1957 when she died unexpectedly, sleeping peacefully in this chair that I'm rocking in now. After that, Grandpa walked his dusty orchard roads every day, shuffling along alone on his worn-out feet. He was blind in one eye by then, but I didn't know why. Grandpa died two decades before Dad moved back into this old house, not long after he and Mom were divorced.

Clemans Farm (1936–1972)

In the early 1850s my maternal great, great, great-grandfather, John Campbell, gave his farm to his oldest son, Marion. Then he led a wagon train across the prairie to Oregon. John and his wife, Lydia, homesteaded with their daughter and their four other boys near Baker City. He was the sheriff of Baker County during most of the Northwest Indian Wars. The main street in Baker still carries his name.

Marion Campbell, the son who stayed behind, took a musketball in the neck at Pea Ridge during the Civil War. It may have saved his life. Released from the Union Army, he returned to his new wife, Susan, and their farm on the Kansas-Nebraska border to recover. They became successful midwestern farmers, had five children, and built a magnificent two-story farmhouse partly, so the story goes, with an inheritance from his father, who'd died in Oregon. Marion and Susan's oldest daughter, Alice, my great-grandmother, married Grant Clemans, the neighbor boy from an adjoining homestead. They eventually inherited the Campbell farm. My Granddad, George, was born there in 1897.

I know that my maternal grandmother, Nina Thomas, also grew up in those plowed-over Kansas prairies. She lived about five miles from the Campbell place, where Granddad was raised. Grandma told me once that she was only four years old when her family moved there from Nebraska, and that she rode the whole dirty way in the back of a creaky covered wagon pulled by two cows. I should know more. How did my grandparents meet? Were they childhood sweethearts or did they find each other some other way? Were they at a barn dance, maybe? Granddad played the base fiddle. Did they meet at a picnic? Share a buggy ride to town? Were they slender or stout then? They married in 1921, that I know, and

I have their wedding picture. She's tall with black hair that barely covers her ears. Her face is round and strong. He's square-jawed and has a straight sandy hairline. His eyes look into you. Neither smile, in the picture. She called him Geo, shortening his name so it sounded a bit like Joe. She always called him that.

George and Nina Clemans started out farming the Campbell homestead with my great-grandparents, Grant and Alice, but as the Great Depression deepened and dust storms blackened the sky their future darkened too. The homestead was in trouble, wheat prices were poor, and the land was mortgaged. The farm could barely support both families so the younger couple—now with a young son, Bob, and daughter, Wanda (my mother)—settled in Agra, a nearby town. Grandma was a good cook; they started up a restaurant.

But as the farmland of the Midwest blew away, they moved too. Uncle Bob and Mom were in high school by this time. The restaurant was doing well enough, but the Campbell place was in deep at the bank and still losing ground. Grant and Alice left first for the fruit country of Washington when the bank foreclosed on their mortgage, taking over the homestead. They left with what they could carry. How did it feel, I wonder, to lose everything and leave behind a history, a heritage, a family, a farm? Were they hurt or angry? Did they feel betrayed—by whom? Or did they simply call it fate, happenstance, bad luck? Maybe it hurt too much to talk about, or maybe they figured it was just the natural consequence of farming and their turn had come up. They never said. George and Nina followed my great-grandparents west a couple of years later. It was 1936.

With money gleaned from the restaurant, Granddad George and Grandma Nina Clemans bought twenty acres of young apples—Winesaps, Romes, and Jonathans—at the edge of Tieton, along a creek and across a wooden bridge at the end of Canal Street. Later, he'd plant a new variety, Red Delicious. Grant and Alice, my great-grandparents, bought a clean, square little house with a white picket fence in town at the corner of Wisconsin and Newland Streets, kitty-corner from the Emanuel Church. Great Granddad Grant shot quail for a restaurant in Yakima and worked the nearby orchards to make ends meet. The Radosevich place was two miles away from Granddad's farm, on the other side of town. The same creek runs through both farms.

Granddad and Grandma's house was painted all white, except for one wall that was covered with brown composite siding creased to look like brick. It still has a broad ivy-covered red-brick chimney leading up from the basement fireplace. Head-high privet hedges that Granddad trimmed square separated their house and flower gardens from the rest of the farm. Each spring they planted the quarter-acre between the creek and hedge to vegetables—beans, peas, potatoes, corn—and berries that Grandma froze or canned and they ate all winter. Granddad fertilized the garden and orchards with manure from his three chicken houses and supplied most of Tieton with fresh eggs. He built on to all our family's homes and taught me how to hold a hammer, drive a nail, frame in a door, and make a straight cut. He showed me how to prune trees in his orchard, and we farmed together for a year in 1965.

I'd like to go back there now, walk through the orchard, sit again in Grandma's warm kitchen and taste her tart apple pies, chocolate chip cookies, fried chicken, sausage and hotcake breakfasts. I could help Granddad candle eggs in the basement early in the morning like I used to. But I can't. Grandma sold their farm in 1972, the same year that Granddad died.

Years later, Grandma gave me a buffalo-skin coat that John Campbell wore when he crossed the plains and while he was sheriff of Baker County. I don't know how Granddad came by his great-grandfather's coat; we don't have relatives left in Baker anymore, at least I don't think so. She'd stored the coat for decades, hanging it carefully in the humid, cool basement of their home in Tieton. I know because I saw it there, in the doorless closet next to the makeshift desk that Granddad used to candle and package eggs.

The coat is knee-length, heavy, and covered with shaggy brown hair that kept out the harsh midwestern and eastern Oregon winds. But its smooth black cloth interior is finely stitched inside with wavy white lines embroidered down its full length. I put it on once, felt its weight, its history on my shoulders. I wondered then about John Campbell and what he was like. Did he wear a beard, carry a rifle? Do I walk like him? I know that I walk like Granddad. Was his voice rough or smooth? He must have been tall. The coat fit. It's stored in my back closet.

The house where I grew up on the Radosevich homeplace is only fifty yards away from this old house with its creaky sloping floors that I'm sitting in now, but I don't go there anymore. Mom and Dad sold it after they divorced. I wonder how it looks inside. What color the walls, linoleum, rugs, counter tops are. Does the fireplace still smoke when stuffed too full of apple wood? Its outside is grown over with fir trees that the new owner planted too close to the house. Its yard is shaded out and the picket fence torn down. Grandpa Joe's garden is packed down and over-stocked with three old car bodies and a broken-down snowmobile. The roof of Grandma Paulina's chicken house fell in last year.

When I come back to Tieton, I often climb the rock butte that looks over this old house and farm. I stand among the sagebrush and boulders and watch the new trees I've planted grow. Or I sit, like I'm doing now in Grandma Paulina's rocker and think, remember. Someday when they're older, I'll teach each of my grandsons how to fish our creek here on the homeplace. How to weave under its willows and cottonwood trees. How to find the deep holes where rainbows lie. We'll hike upstream following the creekbed around the edge of town through acres of well-trimmed orchards until it bends and we come to where Granddad and Grandma Clemans' farm used to be. I can tell them these stories then, about the land they belong to.

Good Wood

Grandma Nina kept her hair short and pulled tight against her head. Her plain housedresses fit loose around her frame. She was a handsome woman and strong but as she aged, she shrank—so much that I'd half-expected her to simply blow away in Tieton's wind one day. She filled her kitchen with the aroma of chicken frying, oatmeal cookies, sausage with eggs, and her apple pies. It was the most comfortable room in her house with hand-built cupboards on every wall and a white Formica countertop that connected a deep double-sink to her electric range. In one corner stood a wooden cabinet that Granddad had made that she filled with her best china and glass souvenirs collected during their infrequent trips to Kansas, her childhood home. A picture window looked across the lawns and hedge into the flower and vegetable gardens. She named every dish she cooked after one of her friends—Viola's scalloped potatoes, Lila's creamed peas, Thelma's fruit salad—but she never followed their recipe, preferring instead to add a touch of this or a pinch of that until the food tasted the way she figured it should.

Now, my grandparent's farm is called Tieton Estates. The year after Granddad died, Grandma sold their place to my second cousin and her husband. They later divorced and sold out to the only developer in the area. Two winters before Grandma passed away, the same year I took over the homeplace from Dad, she asked me to drive her by their old farm. I watched her close as we drove slowly through the soon-to-be subdivision and turned around in the cul-de-sac that had been her garden. She sat upright, straight as a nail, her face rigid, alert, and she said nothing. Their apple orchards had been "dozed"; the trees lay broken and piled, but still unburned. Irrigation pipe, severed and twisted, jutted like bleached bones from the ripped-up dirt. What was in her head, I'd wondered. What was she thinking about—her farm, her husband, her life gone by?

Granddad stood six feet tall and weighed over two hundred pounds. He had a brown spot, a chimera, in one of his gray eyes, trimmed his bushy eyebrows with his Norelco, and always wore blue and white striped coveralls—the kind train engineers wear. He raised chickens, sold eggs, grew apples, and built all of our family's houses.

After Thanksgiving dinner when I was almost nineteen, Granddad found me studying in my bedroom. He seemed to fill the whole doorway when I looked up from my wooden desk. He lit a Philip Morris, waved out the match, and then sucked the smoke deep into his lungs as if it were medicine. The smoke hung around the room after he exhaled. He backed into the conversation. It was his usual way to bring up something difficult, hard, but that he wanted badly to say. He'd been paying Social Security for nearly thirty years, he told me. Then he paused and took another long pull on the cigarette. His words hung in the air like the blue smoke he'd just exhaled. I stayed silent, waiting for him to continue, wondering what that statement was supposed to mean and knowing it meant a lot. It was time he got something back for it, he continued finally. But he'd have to retire, at least part-time, to get his benefits. He'd need someone to help out on his place, a partner. If I wasn't doing anything—he already had the details worked out in his head—he'd make all the expenses, then take out his costs at the end of the year. We'd split whatever was left between us.

We worked side by side in his orchard the rest of that winter, pruning during the afternoons after the trees thawed out. "Don't be puttin' your crop on the ground, lookin' for another one," he grumbled at me one day through the wind, across the Winesap we were both pruning. By watching him make his cuts, I figured out after a while what he meant. There's a balance between one year's crop and the next. Each cut, each decision, has its effect on this year's crop, on the tree's form, and on its ability to produce fruit later. Every cut severs a branch and lays some wood, good wood, on the ground. But each cut also creates the opportunity for more crops later. Prune away too many branches now and this year's crop suffers. But the same cuts force new fruiting wood for following years. It doesn't really matter which branches are saved or cut off so long as balance is attained between the new and older wood, and the cuts are clean. At first he'd walk around my half of each tree, gazing up into it, occasionally taking off a branch or two. After awhile, he didn't bother—apparently I'd got it right.

We could have talked a lot during those frozen wind-blown afternoons, about Kenny, my best friend getting drafted, Tommy Roland blown up in Viet Nam, me being on the waiting list for the Army reserve, the race riots in Alabama, my scholarship to WSU the next year, or Lynn and me. But we didn't. Maybe it was the north wind that made the words freeze up. Maybe I figured he knew, somehow, the way I felt about all those things. Maybe it was enough to just be there, together in his orchard—without words or conditions—working. I looked forward to every day and to each evening sitting in Grandma's warm kitchen, drinking hot coffee around her black Formica table, savoring the fragrances of chicken frying and her apple pie.

Spring that year was warm, and our apple blossoms hurried to break bud. Then in April the Arctic wind started up again. It blew nearly every day but lay calm at night, allowing a deadly freeze to set in. The farmers burned diesel oil in smudge pots, twelve- or fifteen-gallon barrels with chimneys on them that belched thick black smoke and sometimes threw off enough heat to keep the bloom alive during those frigid nights. Usually we didn't light up the pots until two or three in the morning, or sometimes not until dawn, when the wind always dies and the cold air settles in around the trees. One night our frost alarm, set for thirty-three degrees, went off early. It would be a long night. We lit all the pots by ten o'clock. By two in the morning the pots had burned so long that Granddad knew we'd run out of oil by dawn. "I suppose we could use the wood pile," he said. "But I doubt it'll matter. It's just too damn cold out there." His voice chilled too, maybe from too many other cold nights remembered. He hardly ever cussed. "Maybe the wind'll come up a little, warm us up before dawn," but there was an empty unbelieving tone to his voice.

What! Work all winter in that goddamn wind and now just sit here in Grandma's cozy kitchen and watch the pots burn out? Cussing was easy enough for me as I pulled on my nylon jacket and headed out the kitchen door for the woodpile. I know he thought it was futile, my actions rash, even arrogant—to try to warm the sky on such a night—but he helped me anyway. Silently, we loaded the trailer and stuffed the burning pots with next winter's wood supply. It worked. The pots stayed lit. But at dawn the temperature fell anyway, to twenty-six degrees.

Later that morning I walked alone through the orchard. Burnt diesel fuel hung heavy in the air, stinging my eyes. I picked bloom after bloom, pinching them open with my thumbnail to expose the embryo. It only takes a few minutes below freezing to kill an apple blossom, and every embryo was black, frozen, dead. Our crop gone, I baled hops for a farmer in Cowiche that summer instead of finishing out the year with Granddad. Lynn and I enrolled at WSU in the fall and got married a year later. The war was still on and I came up for the Reserve but didn't enlist. That Reserve unit, with some good friends in it, got shot up pretty bad during Tet. Dad rented Granddad's orchard the year after I left for college. After graduation I gave my diploma to Grandma; it was the first in the family. Granddad built a wooden frame for it and she hung it on the wall by her picture window, above the kitchen table.

Ten years after Granddad died, but only a few months before I took over the homeplace, I dreamed that he and I were hunting quail on his farm. We often hunted together during early fall, before harvest-time. In my dream we brushed easily through the wild rye and fescue that grew alongside the creek and willows separating his orchard from the edge of town. It was October again. Apples hung red and dusty from their trees. We wore plaid wool game jackets and gray wide-brimmed felt hats. He was fifty again, the same as I was in the dream. A quail flushed at our feet. It flew frantic across our path, but neither of us shot or even raised a gun. Instead, we walked on with our shotguns cradled under our arms, through his hayfield, along the edge of the orchard, past the garden. We rested on the old wooden bridge that separated his farm from town. "Steve," his words were barely audible. "You ain't puttin' my crop on the ground." His words surprised me. We sat a while longer. Then, he stood and walked slowly up the gravel lane—head down, shoulders stooped slightly to ease his sore back like always—toward his house by himself, without me. I wanted to yell, holler after him to not leave, but didn't. I awoke rested, less troubled about either farm than I'd been for a long time.

I took the dream to mean that there was nothing I could do about his farm and what it had become. Accept it. Learn. Stay balanced. Hold on to the homeplace; like a well-pruned tree, its wood to build from. I'd tried to heat up the sky once years ago. Don't try again; no matter how much I care about him, his place. Now I realize that, like one of Granddad's apple trees, I've been shaped by the events that grew around me, some as harsh as an Arctic wind. But I've also been formed by my own choices; cuts I've made, opportunities taken, some passed over, and knowing when too much good wood is going to the ground.

Starting Again

It's easy to tear out an orchard if you have a ten-ton front-end loader. Just lean its bucket against a tree and give the tractor some gas. You can uproot three or four trees in a row that way if they aren't too big. When the tractor lugs down, back up for a better angle and push them into a burn pile, broken off roots, dirt, and all. The hard part is deciding to do it.

It took over a year for my brother, four sisters, and I to give up completely on the last block of cherry trees on the homeplace. I was the last holdout among the six of us, pleased that the trees, though barely alive, had bloomed at all during the spring of 1995. I'd hoped we could resurrect the six acres with a good hard pruning and plenty of water. Joe, my brother and a fruit inspector for the county, figured we might raise organically grown cherries, if the trees pulled out of it. But as we got into August, the dog days took more limbs, whole trees wilted and died back, and even I realized that we'd better give up and plant again. We contracted with a woodcutter during the winter to take down the orchard. He paid us a dollar a tree for the firewood.

Dad bought each of his small grafted trees from the Columbia & Okanogan Nursery for $1.25 a piece forty years ago. I helped him mark out where to plant each new tree in the old alfalfa field that he'd plowed out the fall before and where, five years earlier, he and Harry Burns, our hired hand, had cut down one of the old Winesap blocks that Grandpa had planted decades before. Harry was older than Dad, and wore thin wire-rimmed glasses that hooked behind the ears. He dressed in faded blue denim overalls that he patched himself, a plaid long-sleeved shirt, and a worn gray felt hat like most farmers did then. He lived alone in one of the one-roomed cabins by the pasture that the apple pickers used when they showed up for harvest each fall. Apple prices had gone sour, 'saps weren't bringing anything, so Dad had wanted to raise cattle and hay for a while. But now cherry prices were better, and there were rumors of a new market for Bings opening up in Japan. He decided to plant again and ordered six hundred cherry trees from the nursery that fall, as soon as the apples were picked. They were delivered one early spring day in 1960, packed together with their roots bundled in wet sawdust and wrapped tight in burlap so only their tops stuck out.

My sister Babi and I were the planting crew, along with Harry. She was eleven and I was fourteen. Harry had already dug several rows of two-foot-deep holes with a nine-inch auger that Dad had borrowed from Hermie Koemple, another farmer who lived nearby. It was mounted on the back of Dad's new 8N Ford tractor. The auger piled up fresh dirt around the holes as it dug. They looked like neat rows of new gopher mounds evenly spaced across the field.

"OK," Dad said on our first day of planting, grinning as he held up one of his new cherry trees by its stem. "The green part always goes up." The seedling was about five feet tall with a few new leaves emerging from stout little branches, but it was bare around the roots except for a few bits of sawdust packing that he brushed away while he talked. He was young then, about forty-five years old, and his brown eyes sparked with intensity under the white "Andy Capp"-style hat he always wore. He bent his wiry, muscular frame over a freshly dug mound of dirt. His knees dug into the soil as he pushed the tree, roots first, to the bottom of the hole. "Now," he said, "Spread the roots out like this." His face was so close to the ground that his voice echoed off the hole's wall. While we peered in, he stood up. "Be sure the biggest limb points this way." He pointed northwest. He wanted the strongest branch to face into the wind. "Fill the hole halfway up with dirt, then tug up on the stem like this," he demonstrated, "so the roots'll grow down. Tamp the soil tight around the roots with this." He handed each of us a round axe handle, blunt on the big end. "Then shovel in more dirt and tamp it in too. After you fill the hole up completely, stomp on it like this." He smiled broadly while half-walking, bouncing, dancing around the little tree. He looked comical and we giggled at his enthusiasm. "Be sure to stomp hard enough to leave a little bowl in the dirt around each tree." He'd turned serious. "Harry or I will come along and pour a bucket of water into it. The water'll seal the soil; it cuts off air to the roots. Air kills young roots."

We planted the orchard exactly the way Dad showed us that first day. It was hard work and we were slow at it, but at the end of each day we were proud of what we'd done. So was he. It took all of our spring break from school.

The morning we intended to start pushing out the old cherry orchard, Polly, my youngest sister, called me at the homeplace. I'm staying with Dad, having arrived late the night before. It's raining and Bill, her husband and driver of the front-end loader, thinks the traction will be too poor to push trees. They'll be up tomorrow when the weather turns better. She doubts if the others will come up either. Carol is still in Spokane. She doesn't know what Joe and his wife, Marian, or our sister Annie are planning to do. "This rain has really fouled things up," she says. "Are you going to stay and work?"

"Sure." I hope I don't sound too disappointed. This little drizzle hardly amounts to a shower in the Willamette Valley where I'm from now, I think as I speak into the phone. I prune my whole vineyard there in the rain every winter. But I say, "Hey, I'm from Oregon, you know, already rusted. I'll windrow the brush the cutter left." Later, after a second cup of bitter instant Folgers with Dad, I walk to my truck and rummage behind the seat until I find my felt hat and rain gear. Damned rain.

The woodcutter has left a hell of a mess of scattered plastic chainsaw-oil containers and cutoff limbs, branches, and whole treetops across the field. He took only the biggest pieces for firewood and left the rest wherever it fell in the snow. I work alone in the drizzle, irritated at all the woodcutter's clutter and annoyed because of our own slow start. Brooding, I trim the brush with the long-handled axe I've brought from home, throwing the cuttings into a long pile forming between two rows of stumps. "This way we can at least push it into a burn pile with the loader tomorrow," I grumble to myself. The work is familiar. I'd often windrowed apple prunings after school when I was a teenager, only I used a twenty-four-inch cornknife and pitchfork back then. Trim off all the branches from any big limbs that Harry or Dad had cut off with the chainsaw. Pile the rest of the prunings into the middle of each tree row with the pitchfork. Dad would rake them into a burn pile with the 8N after all the pruning was done. He sold that apple block to Jimmy Monroe, a neighbor, around 1970 to keep the rest of the farm going. I'd left home for college by then.

God damn it. My axe sinks deep into a downed treetop, cutting it in two. Dad and I laid out this cherry orchard. Babi and I planted it. We weeded around every one of these trees every two weeks for four summers. Even Carol got in on that. When this orchard was nearly grown, twenty feet tall, the freeze of '65 killed half of every tree. The temperature dropped to seventeen below for ten straight days that winter, but there wasn't any snow to insulate the tree trunks from the cold. The bark on every tree split open like a watermelon the next spring when the sap began to flow. I pick up a severed treetop and send the cherry limb, nearly ten feet long, sailing through the air. It crashes noisily on top of the row of brush I'm building.

That must have been the spring that the farm started to go downhill; when Mom and Dad's marriage started to unravel, though I didn't see the split coming for at least another decade. For fifteen years, I know now, they struggled to keep both their marriage and the farm together. I throw another severed treetop, chopped in two, on to the brush pile. By the time I left home Dad was operating the farm with red ink most of the time, and he carried a quart or six-pack of lukewarm Oly or Rainier in his pickup. I lift a whole treetop and heave it toward my growing row of brush. It falls loudly on the pile a short distance away. Beer was about all he drank while irrigating or spraying the orchards during those hot, parched summers. When I went out to help him after my own day's work, he'd ask if I wanted one. There was always a can or bottle of that tepid crap on the seat of the truck, warming up. I usually declined. In fact, I declined so many times that he began to think I didn't like beer, or him, which wasn't true. My axe slices through another treetop and I toss the broken limbs on to the brush pile.

When did my parents fall out of love? Polly, who's twenty years younger than me, asked once, "What it was like, when there was still affection?" The question startled me. I'd assumed that all of us kids knew how much our parents cared for each other and for this farm. "They were happy," I replied. "They both had high expectation, high ideals; they wanted the best for all of us and themselves." But, it must have been different for Polly than for me. Maybe the farm wore my parents down. Maybe it was the other way around. I lift a treetop over my head and toss it toward the pile.

It's late in the morning, near lunchtime, and I'm sitting alone on the wet ground, head on my knees, back against a dead tree stump, facing away from the wind. How did it all fall apart so completely, farm, orchards, marriage? Mom left around 1980, before Polly started high school. She'd retrained herself by then as a nurse and worked for one of the hospitals in Yakima. Those were hard years. Dad couldn't have been easy to live with; neither was she, I suppose. They split up the farm during the divorce, and Mom remarried a few years after that.

How long have I been sitting here, like this? I only intended to stop for a minute, to rest for just a minute before beginning the brush-row again. It's stopped raining, and I hear my brother walking up behind me. He's come to help in spite of the rain. He's ten years younger than me, and had been a pretty decent running back in high school. But I missed all that; I was away at school, then teaching at the university in Davis, California. "It didn't have to be like this, Joe," I say. "Sure it did," he replies while stretching out his arm toward me. He groans a little as he pulls me to my feet. "You know how it was out here. You just weren't around then like Polly, Annie, and I were, that's all." We walk back to the old house together following my row of neatly piled up tree limbs.

Carol is at the old house visiting with Dad as Joe and I come in from the cherry orchard. The drizzle is breaking up. We eat ham sandwiches out on the front porch and drink the cold beers from Dad's refrigerator. We run an inventory out loud of what is left on the homeplace while we eat. Dad listens in, interested for the first time in a long while about anything happening on the farm. For an instant his eyes remind me of the time when he taught Babi and me to plant cherry trees years ago.

"Let's see," I say. "There are just over twenty-six acres left of the original forty, but we have less than half of its water allotment." Dad and Mom sold most of the water rights to make ends meet when farming turned bad. Dad irrigated the orchards by pumping water from a pond he dug in the pasture, but the pond dried up when the irrigation district built a new, less leaky, water-delivery system. They sold the pasture a few years after the pond went dry. "Mom has foreclosed on Lex Newman," Joe says. "Annie says she's going to buy Mom's part of the farm and move back out here. She wants us to work the farm as one place. Did you know that, Steve?" I didn't.

"We'll need to replace the irrigation mainlines for the whole farm then," I respond. Even if the original ones weren't rotting away, the District's new pressurized system would probably blow the old pipes out of the ground. We continue with the inventory. There are some back taxes and penalties to pay. Polly is figuring out how to do that. There's no debt. We still have both tractors on the place, the old 8N and an even older Ferguson, but neither work. There are also two disks, a harrow, and the brushrake. I finish off my sandwich and wash it down with a swallow of beer. "And we have all of us," Carol blurts out, cheerfully.

The rain stops completely by the end of lunch, and the three of us pile the rest of the brush in the now warm, but still ever-present, wind. Polly and Bill drive in with the front-end loader that evening, as we're finishing up. We'll start again in the morning.

N'chi-wanna

The Indians who live along the Columbia River tell about the Frog people, who lived there long ago; how they dammed the river and stood guard over it, keeping the water and fish from everyone else—unless they paid. When Coyote, the mystical trickster/teacher of the tribes, heard what the Frog people were doing, he traveled to where they'd stopped the river, cutting a dentalium shell from a deer bone as he went along. He tricked the Frog people by paying them with the fake shell to drink from the river. When he put his head into the water, he pushed his hands deep into the riverbed, breaching the dam. The water went. The salmon went. And as Coyote arose from the riverbank, he said sternly to the Frog people, "You must not be keeping the water! Respect this river."

Salmon were once so plentiful in the Columbia that the Indian tribes believed the fish consciously let themselves be caught to feed the people. The Indians—Chinook, Wasco, Warm Springs, and Yakama—developed rituals and taboos to assure respect for their Big River, N'chi-wanna, and the fish it provided. They say to fish the river's ebbs, only split a salmon along its back, and never cut it into chunks. Don't waste. Break the taboo, the tribesmen say, and the fish will be too ashamed to return to the river.

It's almost dark and I've turned off Interstate 84 at a truck stop at the intersection of Highway 97 in the Columbia Gorge. I'm on my way to Tieton to plant trees on the homeplace during this long weekend. A row of diesel eighteen-wheelers rumble in the diner's parking lot as I pull in. The truckers leave their engines on while they eat dinner in the nearby diner. A freight train passes by downslope; its engines roar unevenly as its whistle pierces the thin desert air. A grain barge churns its way upstream. This crossroad, I say to myself stepping out of my pickup, has probably always been a busy place; a stopover along the river where Indian traders met to rest and barter. The trails between Canada and California and the Great Plains and Pacific Coast cross here. Wagon ruts of the Oregon Trail cross the grain fields on both sides of Highway 97, only a couple of miles south.

Celilo Falls, flooded now by The Dalles Dam, is only a few miles west of this truckstop, down the Gorge. Coyote, the Indian legends say, taught the people how to catch salmon in the narrow fast-moving chutes of the river, amidst the rapids of those Great Falls. They fished with dip nets that hung loose from a maple hoop lashed to a sturdy sapling pole. When I was small, Granddad bought salmon every year from the Indians who fished the falls at Celilo. I watched them fishing blind, shouldering their long poles against the river's current, from precarious wooden platforms that hung over the rocks at the edge of the Falls. When a salmon hit the net, the fisherman quickly lifted both pole and fish from the water, hand-over-hand. They seemed to float in the mist, fishing in the roar and spray of the rushing water. But that was before the Falls were dynamited, dammed, and flooded over.

Lynn's grandfather owned a farm only a few miles east of here, near where the town of Hanford used to be. He planted an orchard that was irrigated by two artesian wells on a bluff that looked over the river. Her dad, as a small boy, watched from that bluff as Indians dried the salmon they caught along that section of the river, the Hanford Reach. Her grandfather sold his farm to the government when they built the nuclear facility at Hanford, where the first atomic bomb was made. It's ironic; the Hanford Reach is the only stretch along the Columbia that is still free flowing and relatively pristine. The tight security around the nuclear facility suppressed all other kinds of development on that part of the river.

Inside the truck stop, I find a vacant booth in the corner. The table has quarters and dimes embedded in its top and is covered over with thick epoxy that's yellowed by age.

"Ready yet?" It's the waitress. She has rigid, angular features and stringy blond hair that resembles baling twine. Her black denims fit too tight and she wears the tense look of somebody who's lived too long on the edge, places like this river now where indifference is real. It's a look I've grown used to, especially out here on the eastside, east of the Cascades. Maybe it's because the harsh, dry climate makes the consequences of use more noticeable, hard looking.

Salmon canneries flourished for over a century along the Columbia, but now rotting river pilings are all that remain of that all but extinct industry. The canners drove the pilings into the river floor so they jutted out in a long row from a small bay or bend in the river, perpendicular to the current. They stretched jute nets along the pilings, herding the fish into holding pens as they swam upriver. The salmon were then caught barehanded, gutted on the bank, cut up, and canned.

Seines were also used. They were dropped from a rowboat then hauled to shore, full of fish, by horses. Seining was simultaneously banned on both sides of the river in 1947 to protect the salmon runs from total depletion. Even so, few fish return to Neh'i-wana anymore. Two subspecies of sockeye are practically extinct. Nine subspecies of chinook and the entire winter run of coho are threatened, according to the National Marine Fisheries Service. Most years, you can count all the fish in the river through a Plexiglas window at the fish ladder on Bonneville Dam. If you can count them, there aren't enough!

"Ready?" It's the waitress again. Her raspy voice pulls me away from my thoughts. I order the special, "the Big 'un"—a sixteen-ounce cube steak.

Cook's Landing is just downriver from here but on the other side. It was there about forty years ago that David Sohappy, an

Indian fisherman, got caught up in a government sting for poaching salmon. The feds raided the town and arrested most of the fishermen late one night after sixty thousand fish had strangely turned up missing between Bonneville and John Day dams. The fish eventually showed up later, spawning in the Deschutes River—apparently confused by a chemical spill released from a nearby aluminum smelter. But the fishermen were charged with poaching anyway. The judge threw out all the cases except for those against David Sohappy and two others. Sohappy had coincidentally led a successful challenge against the National Marine Fisheries Service for mismanagement of the states' salmon fishery and violation of Indian treaty rights. He was sentenced to five years in prison for selling three salmon to a government agent, but served only two. A U.S. senator was so incensed by the harshness and insensitivity of the sentence that he sponsored a congressional pardon for all three fishermen. David Sohappy returned to the river but never fished again. He died a few years later, physically and mentally broken by the injustice.

The twiney-haired waitress brings my dinner. It's a piece of chopped meat with flour hammered into it, a homemade roll, a scoop of artificial butter, some French fries, and a slab of electric green Jell-O. "Catsup?" she asks.

The dam looked pretty as I drove by it earlier this evening, water cascading across its spillway like a waterfall, bubbly, turbulent. A white mist hung over it; it looked like the Falls used to look. I hadn't seen the river flow like that in more than a decade. It surprised me but then I remembered that the Power Council ordered the spillways opened on all nine Columbia River dams for several weeks this spring, an experiment to flush more smolts downstream. The young fish aren't strong enough to swim to the Pacific without help from the river's current. Flush, that's how the dam engineer described the increased water flow to push the fingerling salmon downstream. I'd hoped he'd misspoken, that he didn't really believe the metaphor he'd used. The newspaper said the amount of water going over the spillways is critical; too much and the smolts die from nitrogen narcosis, too little and the current can't take them to the ocean. I wonder if fish can tell the difference between a waterfall and a spillway as they're swept over it. I hope not.

But the power companies would rather truck fish around the dams than release water downriver. God! So we've come to this, trucking fish because the river can't flow. Several aluminum companies also filed suit this spring against the Power Council. The "lost water," they claim, will increase their electricity rate and raise costs of production. How much does saving the salmon, a species, add to the cost of a roll of aluminum foil? A dime? A quarter? Homeowners like me will have to pay six dollars more a month for electricity. Seventy-two bucks a year to stop extinction. The gas I use driving to the homeplace several times each year costs more. I flew into Portland the other night, on my way back from D.C. There were miles of streetlights on in all directions. What if every other one of those lights was turned off, how much power would that save? How many fish? But the power bureaucrats say it can't be done.

I pick at the food, poke the Jell-O. Why is the death of a species more feasible, more practical, less incomprehensible than turning off half the streetlights in Portland?

There are many reasons why so few salmon return to the Columbia River anymore. Some people blame over-fishing in the ocean and rivers; others accuse miners and loggers of destroying spawning beds along tributaries in the mountains. Still others point their finger at farmers and city people for using too much water. Over a hundred dams now block the Columbia and all its tributaries. All these people talk—miners, manufacturers, loggers, farmers, fishermen, power brokers, politicians, agency bureaucrats responsible for managing the fish, scientists, and Indians. They all talk, argue, and write reports about the fate of salmon in the Columbia and what to do about the loss.

Have you been listening? Salmon have only one way to talk to you, and they only have one story to tell. Their populations fall because of what we do to the river. Maybe the fish can somehow escape our indifference and self-interests; be pardoned. But, how much longer can we go on using this river as if we created it ourselves? Extinction is permanent.

Listen.

Finding Today

A subtle change in light outside the 747 must have awakened me. Is it dawn or dusk, I wonder, peering through the porthole into near darkness at 34,000 feet up? It could be either. I left Sydney that evening while it was still daylight and fell asleep during takeoff, heading back to Oregon. I am flying east into yesterday, toward a distant orange horizon that separates the obsidian sea below from its dome of stars. Is this a sunrise or sunset; is it tomorrow or yesterday? Then, where is today?

I often work in my vineyard alone, but lately, since the weather has gotten better in Oregon, Tyler likes to "help" me prune. He's my grandson, three years old. I park the John Deere between the grape rows when we work there together. I don't need the tractor to prune vines but it gives him something to climb on. We're good company. I hope he remembers this time when he's fifty, no matter where he is, because I want him to grow up feeling the land—this land. To know the good smell of moist dirt, the wonders of worms and weeds, vines, silence, and songs sung only to ourselves. This is what we do, lost among the trellises. I hope he'll remember. I think he might, because I do. Dad taught me to drive a tractor when I was eight, how to thin fruit when I was ten, and to prune trees at thirteen. Before that, I knew the way to every tree in his orchard, and how to find the creek on the homeplace, its paths among the thistle, willow, and cottonwood. I first fished that creek for rainbows with Uncle Al when I was four. They say that early experiences are imprinted indelibly. That you are who you'll become by the time you're three. Today—yesterday—tomorrow.

It is still the same evening as when I left Sydney by the time I arrive at the hospital in Yakima. Dad had gone outside the night before for some reason. Confused in the dark, he'd fallen. He'd awakened a neighbor at dawn, disoriented, bruised, and scared, banging at his door. He's unconscious by the time I arrive, propped up in bed with a tube in each nostril and arm. His chest labors beneath a bleached sheet. His mouth is open, a hollow toothless cavern that craves more air, more air. Leaning close, I smell the organisms on his breath that are killing him, carious, pneumonic. Ashen fingers move ceaselessly across the open collar of his gown. Searching for what? The last sensations of life? I stay with him until after midnight. Susan, my daughter, goes back with me in the morning. He's still asleep. We sit together, watching him, holding hands. What is she thinking? I see me lying there in thirty years—same nose, hairline, forehead, gray stubble around a toothless cave. Does she see me too? We learn so much from our parents. Is this the final lesson? Am I teaching or learning now? My brother and sisters come into the room. When did they leave? That night, I turn on a Mariners game with the sound off and sit alone in the dark with him. When do I ask the unaskable question? Not yet.

By the next weekend, Dad is better. Maybe it's the change in medication, or maybe it is his stout old heart. Still, my brother, Joe, and I board up the windows and doors of the old house. Otherwise, he will want to live on the farm again, alone. Why? Is that land so imprinted into him that it's who he is? It will be the first time in seventy-five years that one of us hasn't lived on the homeplace. He'll live with my sister Babi in town. Just for the winter, we'll say. I'll take his dog.

When we are done, Joe and I walk together through our newest planting of black walnut trees on the farm. It is nearly dark and the air is desert crisp. The yellowed leaves fall from our saplings and we crush the skeletons of summer's last weeds as we brush through them. The ever-present smell of dry earth and sagebrush from the butte nearby hangs stiffly in the air like old incense, surrounding the farm and us. Is this an end or just another beginning, I wonder. I wish I had Tyler with me—on my shoulders, straddle-legged, fingers wrapped tight in his Grandpa's hair, for support. We sometimes walk like that through the vineyard.

Finally, Joe and I stand somberly and silent among the new trees that we planted the spring before, the forest we may never live to harvest, and make plans to plant another.

Part Two

Loss

Now Lynn and I live in the middle of the Willamette Valley alongside a low ridge, between two unnamed creeks that flow year round into the Willamette River a couple of miles away. Fields of grass, sheep pastures, and Charlie Fischer's cherry orchard surround my small farm. It has a vineyard, a little orchard, and a patch of blueberries that we pick all summer. I planted the vineyard the year after we moved in. It has about twenty-three hundred vines and over two miles of wire trellis.

Winter is pruning time. I can prune around a hundred vines in a day if it doesn't rain too hard, but it gets to be routine, even remote after a while. Select four strong canes near the vine's main stem. These make next year's crop. Cut away the rest, clean. String the top two canes in opposite directions along the highest wire of the trellis. Do the same with the other two on the lowest line. Starting at the main stem, count out eight buds along each cane and clip off the rest. Tie each cane to the wire that it runs along with a piece of plastic tape. Then, move to the next vine in the row and do it again. Do it again. Again.

On clear days, I look into three Cascade snowcaps, the Sisters, in the distance and the squared-off patchwork of clearcuts that runs across these mountains. I'm an ecologist, a professor of forestry, crop science, and philosophy. When I prune, I often lose myself among the vines—in the forests and farms I've seen, foresters and farmers I've talked to, and things that make me wonder why they are done. I write these observations, notions, down when I'm back at the house. These stories of loss are about farming and forestry in the Pacific Northwest. They come along with me out of my vineyard.

A Farmer's Neckriddle

It's early spring and I've returned to Tieton for a few days to attend a cousin's wedding and to stake out where the new irrigation mainlines will be located on the homeplace. The last of the brushpiles from the old cherry orchard still smolder. It took my brother, Joe, all winter to burn up the massive piles of tree limbs and trunks that we pushed up the fall before. Once a week during that long cold winter he returned to the farm to roll the smoking trees back on top of each other with a steel pry pole, packing the burning wood to keep the fires going until it was all gone. Now, the land is mostly clear, except for the shoulder-high skeletons of last year's weeds and the sagebrush trying to creep back on to the farm. We'll bury the new irrigation lines later this summer and plant trees again next spring.

Farming, I know, was different when Grandpa owned this land. After he bought the farm over eighty years ago, he planted the first of his orchards. He raised a cow for milk and horses to work his fields. The harnesses hung on the stable walls of the old barn for three decades after he replaced the horses with a tractor. He and Grandma raised four sons and a daughter on this farm. All of my uncles were farmers. Dad stayed on the homeplace, while the others bought orchards of their own. Each one of them sold their farms before they died. Why did all of them but Dad sell off their land, I wonder, as I work this forlorn, open field alone? They either had to sell or go broke, I think, while trampling through the dead head-high stalks of tumbleweed, Jim Hill's mustard, and old sunflower stalks.

And how did Dad manage to hang on for so long? He almost sold the homeplace too, but at the last minute he decided to quitclaim the farm to me. He wanted to keep it in the family for at least one more generation. He told me after we signed the papers that he never truly wanted to sell the farm, only to be out from under its burden of taxes and continual costs to keep a crop growing. Maybe it was poor management or a combination of lost crops, freezes, and bad luck that caused my uncles to sell out, lose their farms. Maybe, but I don't think so; if anything, it's only part of the reason.

During the Middle Ages, in less rational times, I'm told, a condemned man's last chance to avoid the noose was to ask a riddle that could not be solved by either his judge or executioner. Here's a neckriddle that I learned as a boy.

> *A weird creature came to a meeting of men and hauled itself into the high commerce of the wise. It lurched with one eye, two feet, twelve hundred heads, a back, a belly—two hands, arms, shoulders—one neck, two sides. Untwist your mind and say what I mean.*

Too hard? Then help unravel this one.

> *The more it uses, the less it makes.*
> *The less it makes, the more it grows.*
> *The more it grows, the fewer there are. What is it?*

Here's a hint. Today only about one percent of the American population are farmers. And while you're at it, tell me what's so rational about this; it takes nearly six times more energy to grow and market a bushel of corn than that bushel supplies.

My dad and uncles tried to grow more crop even while they worried about the low prices they always got for their apples. Uncle Jack and Uncle Al grew different kinds of fruit, like cherries, apricots, and pears. Uncle Tony left farming entirely and moved to Yakima where there were better jobs. Dad rented more land, which took more of his time, and bought bigger, more costly machines to run the expanding farm. Nothing worked.

Now I know that Dad and my uncles were caught on the treadmill of overproduction and low commodity prices that every farmer in this country has been on for the last one hundred and fifty years. Low crop prices always follow from increased production. It's fundamental rural economics: the more farmers grow, the less money they make for each piece of commodity—a box of apples, bushel of corn, or gallon of milk.

So how do any farmers manage to stay in business? The federal government subsidizes their crops or they borrow, try to boost production, grow more crop per acre, and be more efficient than their neighbors. But growing more is always coupled with the high costs of renting land, purchasing bigger tractors and other equipment, paying for more labor, and buying more fertilizers and pesticides. So farmers almost inevitably end up in debt, sell out, move off their land, and start over in a new profession. The few farmers that remain continue to hope for bumper crops and try to grow still more even though the commodity price of every farm product has stayed about the same for the last fifty years.

When I go back to Tieton, I know where the Marvin farm, the Tasker farm, the Kroger farms, the Monroe and Kazlarich farms, and each of my uncles' farms should be, but they don't exist anymore. The houses these people lived in are still there, but most are vacant. Their land was bought up, first by other farmers, then by corporations that are still consolidating into bigger and bigger companies. The old orchards, trees twenty feet high and planted at fifty per acre, are cut down or pushed out. Some are replaced with new apple varieties. These new trees are genetically selected to grow small so that a thousand or more can be planted on an acre to increase productivity. These little trees, strung on vineyard wire because they are so limber it's impossible for them to stand upright by themselves, barely resemble an orchard. Planting an orchard like that is an impossible expense unless you get the tax breaks of a corporation.

The substitution of genetics, pesticides, fertilizers, and machinery for farmwork was once a good thing for America. In the early part of the last century, these technologies allowed people to leave farms and created the urban workforce needed then to run our mills and factories. Now economists say that urban employment exceeds its demand, while farmers continue to lose their land or their children leave the farm and move to town looking for better jobs. How stable can a way of food production be that drives most of its workers away from the land?

I'm dividing this weedy field into three equal sections, each separated by a parallel line of red surveyor's flags where the irrigation mainlines will be. I mark every twenty-five feet of a heavy steel tape that I've stretched across the field. The noon whistle from the lumber mill five miles away startles me out of my work. That whistle has blasted out the time for as long as I can remember. My job of laying out the mainlines is nearly done, so I decide to eat lunch in Tieton before returning to Yakima for the wedding.

Whenever Lynn and I drive through Tieton one of us says, "Nothing's burned down, so nothing's changed." But that's not true. A lot has changed in Tieton. This little town once supported over twenty businesses and only God knows how many taverns and churches. Now it has one restaurant, a hardware store and grocery. The rest of the buildings are vacant or advertise antiques and rummage sales.

I drive by the house on the corner of Wisconsin and Main streets that was my great-grandparent's home. It's still a clean, pretty little place, painted white with a black asphalt shake roof and surrounded by a well-kept, whitewashed picket fence. I've seen the Hispanic couple who lives there now. By the playground fixtures, trikes, and other toys scattered around the yard behind their picket fence, I figure they must have at least two kids. Grandma told me that his name is Pedro and that he works for one of the big apple companies as an irrigator, sprayman, and pruner. He will never own the land he farms. People like Pedro and his family are another consequence of the farmer's neckriddle all across America.

I park in front of the only restaurant left in Tieton, kitty-corner from where Newland's Pharmacy used to be, and go inside. The waitress is about my mother's age. Her hair is dyed bright red. She is the widow of a logger who lived a few doors away from my great-grandparent's home. I used to visit her with Grandma when I was little. The jobs in Tieton are so intertwined. While I was growing up my best friend's father was a logger as well as a farmer. He worked for the lumber mill in the next town for over twenty-five years while he tried to keep his own farm from going under.

The waitress's name is Louise. She doesn't recognize me as I sit down at one of the lopsided tables and order a hamburger and cup of split pea soup—the daily special. I don't say who I am. There's no point. I absentmindedly read the bulletin board that's on the wall across the table from me. Half of the bulletins are written in Spanish. There's a dance at one of the churches tomorrow night.

The spring wind is frigid outside. "It'll probably freeze again tonight," Louise says as she sets my soup on the rickety table. It steams deliciously. I think back forty years to when Granddad and I farmed his place on the other side of town for a year. That crop froze out. It only takes a few minutes of cold to kill an apple blossom. If it hadn't been for that one cold night, I might still be farming in Tieton, I think, as I taste the thick warm soup.

Louise's off-handed comment has me reflecting again about my place in the neckriddle. It wasn't just a cold night in 1965 that caused me to leave home and go to college. Even back then I felt the farmer's neckriddle at my shoulder, but didn't know what to make of it. I was like one of Granddad's apple trees being pruned that winter—the first cut results in other decisions until the tree is shaped. I transfered to WSU after the freeze and passed up homesteading on the Royal Slope of the Columbia Basin, the last Homestead Act in the country, after graduation. I also turned down a county agent's position in Wenatchee that summer. These opportunities were left on the ground like good wood cut from a well-pruned apple tree. Instead, I enrolled in graduate school at OSU and four years later started teaching at UC, Davis. It had only been seven years since Granddad and I froze out.

During my last year in grad school, Mom asked if I'd return to Tieton and run both the homeplace and Granddad's farm. "The orchards are old," she'd said, "but the trees are healthy. The farms need new blood." She figured I knew how to revitalize the orchards so the farms would stay afloat. But I didn't. I wanted to take her up on the offer, but declined; farming by that time scared me. I knew the neckriddle but not how to solve it. I had my own growing family to support too and those two farms, it seemed to me, could barely support the people who lived on them then. Lynn, our three kids, and I stayed in California for over twelve years. I learned about botany, ecology, and forestry, and taught integrated pest management.

Now I'm a farmer again; I planted a vineyard two years ago. But I'm a scientist too, a professor of ecology, and have studied weeds for most of my academic life; first only to show how to kill them, then to understand why they reduce crop yields in farmers' fields and decrease tree growth in forest clearcuts. But weeds still grow everywhere, and foresters make more and more clearcuts even as they learn to plant and grow trees in them better. Now I wonder if my research has helped at all, or if it is simply another short phrase in the continuing neckriddle that asks why farmers lose their land and move away from towns like Tieton.

About three decades ago, farmers bothered by the prevalence of weeds in their fields—not to mention their declining profits—turned to herbicides to kill these unwanted plants, and to grow more crop. But that "solution" only drove down market prices once again and increased production costs. It also created other problems like soil erosion and chemical contamination of food, water, and soil. And all the while farmers and loggers moved away from rural towns as large companies bought up their land, getting bigger and bigger.

Now, other agricultural scientists have a new variation of the same tired technology to kill weeds. They splice the genes of unrelated organisms into crop plants to make the crops resistant to herbicides. That way, growers can increase yields by spraying more times or spraying more effective chemicals without the fear of injuring their crops. But even if this newest approach to weed control is successful and crop productivity improves, farmers will still be faced with the increased cost of buying herbicide-resistant seed and, of course, the chemical to spray year after year. And the nagging problem of overproduction followed by low market prices, more displaced rural people, and the consolidation of farms will continue. It is a special insanity to expect a different outcome when approaching the same problem in the same way.

Neckriddles are necessarily obscure. They intentionally deceive, befuddle, but become clear once the answer is known—like in the riddle of the one-eyed garlic salesman that I learned as a boy. So, is the solution to the farmer's neckriddle still too difficult to see? Then look at the problem of overproduction, low market prices, and displaced farmers in a different way. Imagine a scenario where farmers grow food based on an understanding and acceptance of nature. One where farmers are accountable to their land, adapting their tools to the way plants and animals grow, not the other way around. Weeds, insect pests, diseases, and low fertility would probably not even exist under such a scenario because they would simply be incorporated into the normal cycles of production—growth, reproduction and regeneration, predation, competition, facilitation, and herbivory.

I know a farmer who grows raisins in Fresno. When I first met him, he'd just returned from Viet Nam and was losing the family farm to debt, high production costs, and low market prices. Neither I nor other scientists then could help save his farm. We didn't know how. Eventually he figured out how to lower production costs by not borrowing, relearning how his grandfather had farmed early in the last century, and modernizing that old equipment using hydraulics and common sense. He decreased the load of pesticides, herbicides, and fertilizers in his vineyard by over half while his yields stayed the same.

Another farmer who lives in Iowa figured out how to grow corn without any herbicides, and to use the mulch from tilled up weeds as an organic fertilizer. Because he saves moisture and plants his corn later than his neighbors when the soil temperature is warm, his corn grows faster and he avoids most serious disease and insect damage to his crop. Both these farmers market their crops in conventional, not organic, markets, and both show a profit each year by reducing their costs, not by increasing their yields. But barring a sudden shift in awareness, most farmers seem firmly stumped by their own neckriddle.

So who am I in this neckriddle and who are you? Are we prisoner, judge, or executioner? Can we solve the interwoven problems of growing food, overproduction, low crop prices, technological dependence, and farmer displacement, or are we the problem? Farmers, scientists, consumers, untwist your minds and help me think broadly about this riddle because it is impossible to solve using the same way of thinking that created it. And I fear that the continued existence of some large land-holding corporations and the manufacturers, suppliers, and wholesalers of farm supplies depend on our *not* figuring it out.

Consumers, people like you and me, want cheap, high-quality food to eat. We also want it without contaminants, and without harm to our environment or to the people who work the land. Do you know where your groceries come from, how your food is grown? What if our wishes about how food is grown were better known, less contradictory, more direct? So, I must observe who benefits from my habits and tastes and change them when I don't like what I see.

Scientists, the sad fact is that almost every innovation in agriculture for the last one hundred and fifty years has driven farmers from their land, caused environmental contamination, or eroded the soil. Be aware that science, as a way to solve agricultural problems, is inextricably bound to the markets of food production; that research now only counteracts problems generated from older production practices. This research does not improve the well-being of farmers, foresters, or their land. So if my work as a scientist enhances only a few rich and powerful people or companies, if it causes self-sufficient people to lose their land or livelihood, or if it degrades the environment, then I must work to find solutions that don't.

Farmers, we are complicit in our own neckriddle by using more and more chemicals, machinery, and energy each year. While there seems to be little choice, alternatives to our riddle exist. We must weigh carefully the technology of how to grow food. The treadmill of chemicals, energy, and machines now has supplanted almost completely our real work—knowing about the land and its biological processes to grow food or wood.

I pass the homeplace on my way out of town, but don't stop. If I do, I'll be late for my cousin's wedding. The red surveyor flags that mark where the new irrigation mainlines will be buried stand out clearly among the weeds. I know about the farmer's neckriddle now, saw it work first hand and found its solution. When we farmers replace our knowledge of the land with too many subsidies of genetics, chemicals, energy, and machines, we replace ourselves.

This old farm is ours, my brother's, sisters', and mine. I know that now. We'll plant again next year.

Mad Cows

It's been snowing for three days and I'm peeved. I can't get back to pruning my vineyard until this storm breaks and the snow melts off. Right now, I can barely make out my neighbor's cow herd on the hillside a few hundred yards away through the sheets of sideways-blowing snow. The cows stand in a semicircle around a broken bale of alfalfa hay, butts facing north, tails against the wind. The snow blows across their wide black backs. It accumulates in their fur.

Yesterday the storm did break for a few hours but after the temporary thaw the snow froze again overnight, leaving an icy crust of white across the hillside. The cows trudged awkwardly across the pasture this morning, each more or less in a separate direction. The crust broke under the weight of each step. Nostrils vented steam into the crisp air as each cow deliberately pawed through the broken ice and snow until a brown tuft of grass lay exposed. Then she poked her nose into the crushed powder, wrapped her warm tongue around the grass and pulled the clump into her mouth. But now, in the midst of still another storm, they wait with feigned patience for the farmer to come. He needs to break the ice that covers over their water trough. One heifer restlessly pushes the broken bale with her nose and stomps it into the snow with her hooves. They know he'll come, but the whole herd is ticked off. I can tell.

I don't know what these cows have to be so hot under the collar about. They're breeders. Life could be a lot worse, like the calves they'll drop this spring that will be auctioned off as yearlings before next winter's snow. Those yearlings would have something to grumble about if they knew their fate as hamburger or beef steak, but they don't. The neighbor who owns the pasture on the other side of my vineyard buys a couple of calves every spring. My four-year-old grandson, Timmy, named them last year. He calls one Blackie, the other, Brownie. They went away last fall but came back again in the spring, or so it seemed to Tim. Only Brownie didn't come back, and Blackie, who was smaller than before he left, brought a new friend—one with a red body and white face. Timmy asked me last summer where Brownie is, why she didn't come back to the pasture with Blackie, but I didn't tell him. He'll find out soon enough, I figured. He named the new calf Whitey, and I got a reprieve from the question for at least another year.

It's not that I'm squeamish about knowing where my meat comes from. I grew up on a farm and butchering was one of the fall-time chores, a yearly task that needed to be done to keep meat on the table. We had two milk cows, Bess and Red, and each spring both dropped a calf—once in a while twins, if we were lucky. Dad raised all his cows on grass for most of the year, but during winter, after snow covered the dormant pasture, he fed them alfalfa hay. He sold all the yearlings in the fall, except one. That one he raised for one more year and butchered it the following fall. He never let me name the calves. You can't give your dinner a name.

Granddad always helped Dad butcher. So did I, when I got old enough—around thirteen or fourteen years old. The job went like this. Dad would call Red and Bess to the barn, where he'd lock them inside. Then Granddad would walk into the pasture holding his twenty-two pistol behind his back, usually right up to the unsuspecting two-year-old. Pop. One bullet in the head and it was done. Then he'd bleed the animal and drag it from the field with a tractor. We hung the carcass upside down from a tripod to skin, gut, and cut it up. There's a lot of meat on a cow other than steaks, roasts, and hamburger—heart, liver, tongue, tenderloin—and my family ate it all. Dad and Granddad loved brains. They scrambled them with eggs.

Now, let me tell you about something that really gets me boiling. Two weeks ago the Department of Agriculture announced that Mad Cow Disease was found on a dairy farm in eastern Washington, not far from our homeplace. That news is alarming enough, because the disease (technically, bovine spongiform encephalopathy) can be transferred to people who eat infected meat. The illness is called Creutzfeldt-Jakob syndrome in humans. It's fatal, as it is in cattle. USDA officials say the meat from the infected cow could have been shipped to as many as eight other states. What burns my tail is that they try to reassure us by saying that the risk is small, that only the brain, spinal matter, and lower intestine carry the brain-wasting illness—parts of the infected animal that are unlikely to be eaten by people. Bullshit!

I knew a man who died from Mad Cow Disease. He was a neighbor who lived in our old subdivision. He had traveled to Britain in the early 1980s and was there when the first outbreaks of the illness were discovered. He was one of the 143 people who died of the disease then. It takes up to five years to incubate in a person's brain, then wastes away the entire nervous system. My neighbor lost all memory, muscle, and other body functions as he withered away for another half-decade. Maybe he was fond of cow brain, spinal matter, and lower intestine, but I doubt it. He must have contracted the disease from infected nerves in the meat he'd eaten. And, in spite of all the assurances, food safety officials have recalled over ten thousand pounds of meat they think may have been mixed with that of the infected cow. It takes less than a gram of infected tissue to infect a cow. How much to affect a person is still uncertain.

If that's not enough to bristle your feathers, consider how the infection gets passed from cow to cow and into our food. Cows transfer the virus by eating infected meat! How can that be? Cows are herbivores. Cows eat grass, leaves, foliage, vegetation—not other cows. Apparently the fate of old, diseased, or dying cows is to be ground up, pulverized, and fed as protein supplement to—

get this—other cows. It's an economic thing. It's cheaper to feed ground-up dead cow to keep a herd productive than to buy soybean meal or alfalfa hay. Occasionally, one with bovine spongiform encephalopathy gets mixed in. It's a risk we're all supposed to take, especially if we don't know about it.

Feeding cow-derived protein to other cows is now prohibited in the U.S., so now some farmers feed the ground-up cow meal to chickens. Then, the chicken droppings, nesting straw, feathers, whatever, are pulverized and sold as—you guessed it—protein supplement for cows. Have these farmers lost their minds? Cows don't eat cows—not as steaks, hamburger, pellets, or as a powder. They don't eat cow with catsup, mustard, salsa, or as granulated chicken manure. Chickens don't eat dead cow either. These farmers could run their businesses based on common decency, common sense—in spite of their crazy economics, not because of them. They could graze their cattle on grass and feed them alfalfa for protein like Dad did. They could fertilize their fields with chicken manure. But that would be foolish, insane.

No wonder the cows are mad.

Clearcut

I'm back in my vineyard after the snowstorm. I've pruned almost thirteen hundred vines so far this winter. The snow on the mountains makes the clearcuts stand out, white rectangles purposefully cut into a sea of green. They make me think of Charlie Wakenshaw.

I'd arrived early, to get the lay of the land, before meeting Charlie Wakenshaw at the Siletz Little Chief café. He'd called me a few days earlier wanting to show someone, anyone, what was happening to the forest where he lives. I agreed to meet him in Siletz, a logging town in the Oregon Coast Range, uncertain about how I might help. As I parked my truck, I saw two men through the café window, one waving a plastic coffee mug in the air. They were Indians and dressed like loggers. Their bodies said they were angry but probably not with each other. I warily entered the café, still hoping to check the place out, but the logger holding the mug turned and walked purposefully toward me before the waitress could even bring coffee. He stopped at my table, his voice coarse, harsh, husky. "Hi, I'm Charlie," he said, grinning, holding out his huge right hand. The pleasantness in his greeting surprised me. Was I the fellow from OSU he'd phoned the other day, he asked while slipping easily into the seat across from me, obviously at home in the Little Chief.

Charlie wore a gray felt hat, the ten-gallon kind except with a short, stout brim—the type you see in old photos of Indians standing by cigar stores. His hatband was made of pale leather with a delicate Native American design of small turquoise and white beads stitched on to it. Half his right index finger was missing, cut off below the first knuckle. He was clean shaven, except for a thick black mustache. Several teeth were missing on the left side of his face and that made his lower jaw move sideways, causing some of his words to slur as he spoke. He wore a red kerchief bandana rolled tight inside his hat. "I'm glad you're here. You gotta see what they're doin' to the mountain. Christ. It's slicked off clear to Corvallis," he grumbled, waving his hands more or less eastward—referring to the forty miles of Coast Range I'd just driven through.

We left the café and soon were winding along a one-lane road that followed the river. It was surprisingly well constructed, reinforced with layers of crushed rock and gravel, expensive, built to last. Every few miles we faced a loaded log truck head-on, going downhill with its air brakes rapping, and quickly turned out to let it pass. The road switched back to a ridge where it connected into a labyrinth of similar roads all across the Coast Range. I knew that I should see trees with deep furrowed bark; ancient Douglas-fir, red cedar, hemlock—some enormous when Columbus sailed. Ten of me, arms outstretched, couldn't reach around any of them. Their heads would be lost in the low clouds of the mountains, their branches draped in gray-green lichens—old man's beard. There should be bigleaf maple and alder, vine maple, salmonberry, green huckleberry, rhododendron, and salal, hazel, sword fern, deer fern, all the way down to the clover-leafed wood sorrel on the soil's surface. And the earth should be covered with duff, decayed needles and leaves, rich soil created by millions of miles of fungal mycelia that cycle the forest's nutrients.

But I didn't see any of this. What I saw were clearcuts from river bottom to ridge top, and in the cuts everything was either gone, broken, or piled up. First, the best trees had been taken to the mill. Then, everything else had been burned, scarified, or poisoned; firs, cedars, maples, ferns, moss, fungi, lichens, and anything else that depends on these things to grow. What remained were re-planted

Douglas-fir, burned-out slash piles or the charred earth of broadcast burns, fields of yellow-headed Scotch broom, thorny salmonberry as tall as a man, or thickets of alder too dense to walk through. Released from the overstory trees, these plants would stay around for decades, impenetrable unless sprayed with herbicides to suppress their growth.

After several miles along the ridge road, Charlie suddenly stopped his truck at the bottom of a small ravine. We were surrounded on either side by steep muddy slopes, piles of slash, and stumps still in the ground. He scrambled noisily out of the truck, and crashed through the brush to the base of a stump over four feet across. It was cut high and still bleeding. It was an old tree; the kind loggers call hooters because their limbs grow thick and near the ground. Ruffed grouse and owls, hooters, sit in trees like these, using the branches for perches and cover. The wood in hooter trees is too knotty for good lumber, so loggers usually leave them standing unless the price of timber goes up.

"Ohhh no," Charlie wailed in disbelief, caressing the cut surface of the ancient tree with his palms, touching the stump like the face of a lost lover. "They cut down the elk-tree! Elk always come here, to this tree, when it rains. Granddad brought me here before I could even walk to see 'em." He looked at me helpless, staggered, shaken—as though his explanation should somehow bring the tree back.

"Listen, I ain't no goddamned environmentalist, " Charlie growled later, back in the truck, "but something's gotta be done out here. These goddamn clearcuts ain't got nothin' in 'em, 'cept stickers and broom. The tribe can't do nothin' and the state won't do nothin' about what them timber companies are doin'. Sure as hell, I've been a logger most of my forty years, but I ain't never seen nothin' like this. I ain't against cuttin' a few trees. Hell, Granddad did it. But pretty soon, there won't be no more life left out here."

Three Lies

Fool me once, shame on you. Fool me twice, shame on me
—old German proverb

I came across a magazine article several years ago while rummaging through a knee-high stack of papers that sits perennially near my computer at the university. Professors are noted for these caches of old scientific journals, newspaper articles, sometimes entire magazines that lie neglected in the corners and crannies of their offices; stuff always interesting enough to save and read more closely, only later. This paper caught my attention because of its yellowed torn pages, some with the corners bent down. I don't know who gave me the article or even how it got into my stack. *Fortune, Feb. 1945, 31:169-175* was written across the top of the first page. At the bottom, a note in the same hand said, "Steve, this gets at the problem of sustained yield."

The article used a Forest Service inventory made in 1938 to assess the status of Northwest forests for its readers back then, people with money, investors—same as now. The article concluded that the region's private forests were in such decline from over-harvesting and wildfires that loggers had to slow down so that tree growth could catch up with the cut. It predicted that if they didn't Washington and Oregon would run out of corporate-owned old-growth and second-growth forests by 1962; that a new source of logs had to be found. *Fortune* also surmised that the Forest Service would eventually bow under to the timber lobby and add trees from our national forests to the base of timberland, which would extend the cut to 1999—thirty-seven more years. *Fortune's* projections came true. Now less than three percent of our aboriginal forest remains and that only because it is so remote as to be inaccessible to the saw.

All foresters now swear by the concept of sustained yield. It's a fundamental principle taught in every beginning forestry class: the amount of wood being harvested from any given stand of trees in a forest must not exceed the amount of growth put on by the whole forest during that harvest period. For example, if a forest is one hundred acres in size, then it could be divided into a hundred one-acre stands and each acre cut, one year at a time, beginning with stand 1 and ending with stand 100. Thus after one hundred years, stand 1 could once again be harvested, maintaining a sustained yield from the forest for one hundred years. When foresters practice sustained yield, theoretically, the cut could rotate perpetually across the forest landscape from one stand of trees to the next, to the next, forever. That's the principle.

But in the Northwest, because of the historically high industrial harvest rate and the abundance of national forests still left to cut, the practicality of the concept came to mean little more than using Forest Service trees to keep corporate mills going. Industrial forestland by itself was simply not vast enough to sustain the magnitude of the cut; the remaining trees couldn't grow fast enough to replace the amount of timber being taken out. Since there wasn't enough time for the forest to regrow between cuts, the timber industry would either move away or collapse unless national forests were added to the timber-base.

There's the first deception, that foresters have been practicing sustained yield in this region for more than a century. Instead, I think a grand era of exploitation just ended—a strategy or a tragedy, depending on your point of view, that led to the eventual replacement of our aboriginal forest by the industrial tree farm. But the replacement is only a facsimile of the real thing. A tree farm is about as similar to a forest as a wheat field is to a prairie.

Prairies are more than grass; forests are much more than trees. Prairies and forests are webs of organisms interwoven by coexistence and mutual dependence. This mutuality, this reciprocity, assures long-term productivity and sustainability of the land. We may not be running out of trees here in the Northwest, but we are running out of forests.

All forests in this region need some kind of disturbance to open up their canopy of tree cover, which allows them to regenerate themselves and stay productive and healthy. In nature, regeneration usually happens in mature forests when an old tree falls down, uprooting the soil, making a spot of sun for new seedlings or saplings to grow tall. Some forests need larger, more catastrophic disturbances, like fires, for them to regenerate well. So, here's the second lie—that forests in the Northwest need clearcuts to regenerate themselves. Foresters here learn that forestry begins with a timber harvest because a clearcut disturbs the forest. But clearcuts are neither the only nor the best kind of forest disturbance; they're just the cheapest way to take logs out of the woods and to the mill. They hardly resemble natural disturbances at all. During a clearcut most of the good wood of the forest is cut down and hauled away. But even after a wildfire, logs stay, standing or lying on the ground, often for centuries. As the wood decomposes, the forest's nutrients cycle back through the earth and into the new trees that establish. But no! Timbermen here will tell you, with all the conviction of the bottom line, that a clearcut is the first step in forest regeneration.

There are biological reasons for the way forests grow. These relationships between the trees and the environment they're growing in establish the minimum size that openings in forests can be for natural regeneration to happen. For the Douglas-fir forests of western Oregon and Washington, that opening is about three to five acres—a far cry from the gaping two-hundred-acre clearcuts of the last half-century. When clearcuts are used to regenerate forests, the same biological relationships set the length of time needed between harvests to sustain maximum timber production. This rotation time from one harvest to the next varies by location of the forest land, but in Oregon and Washington it's around one hundred years. That means only about one percent of the Douglas-fir forest here should be cut annually if the region's wood supply is to be sustained. Unfortunately, economic interests usually get in the way of biological principles. The rotation time shortens and the forest loses out.

The paradox is that as rotations shorten, timber production falls. So, here's the third deception—that short rotations increase timber yield. Bull! If a one-hundred-year rotation is cut in half to fifty years, as most timber companies do now, timber production will not increase. It will fall by as much as four-fold. In other words, the forest will yield about a quarter of the wood when fifty years old that it would yield at one hundred years. Why? Because big trees always put on more wood than little ones. It's like a bank account; the trees in the forest are the principal and their growth rate is the interest. A large principal "invested," even at a low interest rate, will yield more than a small principal invested at much higher rates because there is so much more principal on which to collect returns. It's the same with a forest. Clearcutting perpetually replaces the largest "principal" of the forest with little trees. Young trees often have higher growth rates than older, bigger ones, but it's not fast enough to make up for the large loss of principal. So shorter and shorter rotations can only result in less wood.

Plant more trees, the timbermen say. But increased planting is not the solution to the paradox of short rotations, because the number of trees needed to restore the yield lost is more than the land can hold. When the one-hundred-year rotation is cut in half, foresters would need to plant thousands of trees on every acre of clearcut to restore the lost growth, not the few hundred they do now. Trees can't grow that close together and become big enough to harvest, even if it were economically feasible to plant so many of them. Then fertilize and spray the clearcuts, they say, manage the seedling trees intensively like farmers do their crops. But farmers fertilize and spray crops annually to improve their yields, expenses that make intensive management economically untenable in forests and again force foresters to short rotations and lower yields.

So we've been fooled, deceived, hoodwinked by the lumbermen, perhaps deliberately, perhaps not. The question now is, is there another way? Are there alternatives to ever-lower timber yields, a declining forest industry, and a perpetually rotating clearcut across our landscape? Ironically, the solution to the dilemma of timber production here is to slow down cutting and practice longer rotations, not shorter. Or to have no rotation at all by logging only certain trees or small patches of them and letting the rest grow and regenerate naturally.

Several years ago, again while rummaging through my stacks of lost papers, I found a "cruise" sheet for a tract of cut-over land, a second-growth forest owned by a family I'll call the Bakers. Timber cruises are how foresters estimate the amount of wood in a forest stand they intend to cut. The tract had been clearcut fifty or sixty years before, and the trees when the Bakers bought it ranged in size from saplings to some over a hundred feet tall. The first cruise on the Baker family land was conducted in August 1967. It showed that the stand contained about two million board feet of lumber. The Bakers selectively harvested their land five times over the next twenty-one years, eventually removing 4.2 million board feet. In 1988, after their last harvest, the land still held 2.2 million board feet of timber—more wood than had been there before the logging! But what if the Bakers had clearcut their land back in 1967? What would their yield have been? Something less than their original two million board feet, no doubt; harvests never totally recover all of the wood cut down. Plus whatever they might grow from a new planting of trees in twenty-one years, which is several hundred times less than the 4.2 million board feet that the Baker family collected.

Many private forest owners, like the Bakers, now would like to grow their timber on longer, not shorter, rotations or to selectively log and not clearcut at all. They know they can produce more wood when the logs are bigger. But with only one or two exceptions, the forest companies that practice short-rotation forestry also own the sawmills. These mills only accept small-dimension logs and cull or downgrade the big ones to a lower price, forcing private forest owners to grow small trees in order to sell any timber at all.

Timbermen say their system of production isn't broken, that planted forests grow back, that with management their plantations stay productive. True enough, I guess. But only if foresters can somehow stay inside the strictest bounds of sustained yield; if they, like the Baker family, can let the forest and not a company balance sheet determine how long the rotation should be. Still, I wonder as foresters embrace shorter and shorter rotations and as their legacy of clearcuts perpetually roams across our landscape, if foresters haven't deceived themselves worst of all.

Inventory

It's raining in my vineyard this morning; another cool Oregon drizzle that blew in from the west across the Coast Range to my back. I'm warm enough beneath my weatherworn raincoat, broad-brimmed hat, and new fleece jacket—a Christmas gift from my oldest son and daughter-in-law. I've pruned almost two-thirds of the vineyard and thrown the canes I've cut off, last year's good wood, on the ground between the vine rows. The trellis seems bare, open, exposed, except for the four canes I leave behind to set the next crop. Pruning revitalizes the vines after harvest.

I always look east into the Cascades, across the valley, when I prune. There's a clearcut up there, probably fifty miles away, noticeable from here because it gets bigger each year. This winter, it's merged with several other cuts on the mountain. Their straight lines and rigid square corners are even more visible today because of the new snow that fell in the mountains overnight.

Foresters say that clearcuts revitalize the forest for another crop of trees. I'm not so sure.

It was early morning several summers ago and already hot in the central Oregon Cascades. Will, Al, and I sat on the tailgate of their half-ton truck and laced up our high-topped leather boots. Will and Al are company foresters; they agreed to let me join them for a few days as part of a research project on clearcuts.

Both men wore loose-fitting black jeans, red suspenders, and T-shirts beneath long-sleeved cotton shirts. They leave the shirt front open but turn up its collar to keep fir needles from falling down their necks. The sleeves are buttoned down tight around their wrists to protect their arms from blackberry thorns. We put on fluorescent orange hardhats and surveyor's vests and cram the pockets with a quart bottle of water, tree diameter measuring tape, marking paint and metal tags, tree flagging, and a compass. Al carries a telescoping height-measuring pole. Will slings a small battery-operated computer, a datalogger, over his shoulder as we walk toward the site we'll measure today.

They have been out all summer and will download today's tree inventory over the phone to the main computer at company headquarters tonight. Most of the trees in the clearcuts they've been measuring were planted ten to thirty-five years ago. Most are on steep slopes—not exactly cliffs, but you don't want to fall off one. Today, the ground is "flat," which means you won't roll downhill if you fall. Will pulls a surveyor's map from his vest. "We'll start here," he says to Al and me as he examines the map he's carrying. "Site GP/OCR-4568," then he closes the map and tucks it into a vest pocket.

Al, in the lead, pushes purposefully into the planted clearcut through the head-high Scotch broom and blackberry for about twenty yards. Then, under a continuous rain of dry fir needles, the men flag out a sample plot. The plot is always the same size, a square chain, sixty-six feet on a side. The sampling is systematic and soon it becomes monotonous, measuring every tree in the plot, every tree, every tree. The uniformity inside the stand makes me dizzy. The trees blur together; all are equally spaced, each with identical sets of branches, symmetrical whorls that overlap, touching each other at their trunks. Every tree is a Douglas-fir. Every tree is thirty to thirty-three feet tall. Every tree is 6.0 to 6.6 inches in diameter, and they are planted in rows as straight as Nebraska corn. The understory is always some kind of sticker—blackberry, salmonberry, or thimbleberry—if it's there at all. The brush is sprayed with herbicide when the trees are little, then the saplings shade out most of the other vegetation as their canopies grow together.

A rain of dead fir needles drops steadily as we move methodically from tree to tree. It's a sign of tree stress, probably for water. It's humid, stuffy inside the stand. I quickly drain the bottle of water I'm carrying. I feel disoriented. There are no landmarks, no distinguishing features, no differences; only identical rows of identical trees. We marked our trail so we can find our way out.

After a clearcut, Caterpillar tractors pile limbs and trees too small or damaged to take to the mill into a slash pile to be burned. But if the slopes are too steep for bulldozers, the slash is left where it falls and the entire area is burned later. Then, planters can move more quickly through the clearcut, planting row after row of seedling Douglas-fir ten feet apart. Brush grows back quickly after a clearcut, so planters need to be in the cutover soon after the trees have been cut down and trucked to the mill. Otherwise, stickers, broom, or alder choke out the seedling trees they've planted. Every cutover is sprayed one to three times with herbicide to suppress the brush. It's always the same procedure. Foresters call it a prescription; cut the trees and pull them to a road, pile the slash, burn it, plant new trees, spray the brush, spray it again, and wait for the little trees to grow. Then cut and do it again, then again, again.

When my youngest son, Matthew, was a teenager, he caught walking pneumonia three years in a row. Our doctor prescribed penicillin each spring that Matt got sick, which cured him every time. But after the third time, I insisted that Matt wear his winter jacket for a month longer during spring. He did so reluctantly; the pneumonia never came back. I think that spraying herbicide over clearcuts is similar. Brush is a symptom of what happens every time a forest is cut down. Herbicides are like the penicillin the doctor prescribed to Matt. It temporarily cured the disease, but it didn't stop him from catching pneumonia each spring. His change in behavior did that. Foresters spray brush in their clearcuts year after year. How long will it take for them to recognize the real problem and change their behavior? Having a herbicide to cure the symptom of brush in clearcuts is not a good reason to keep doing the things that cause the problem in the first place.

The three of us take a different path out of the planted clearcut we've just measured, walking toward a patch of naturally regenerated second-growth trees that Will has located on his map. Along the way, we pass through another tree stand, GP/OCR-4570, where all the alder saplings in the thicket have been slashed down and sprayed out so the smaller planted Douglas-fir seedlings can grow better. Native plants like alder and salmonberry are weeds to foresters because they take over clearcuts quickly after the logging and over-top their newly planted Douglas-fir seedlings. But saplings that grow from alder seeds dispersed on to disturbed ground help to keep the soil from eroding away. Shrubs like salmonberry reestablish from preexisting root crowns and do the same thing.

We continue on and arrive at the edge of the second-growth stand that we've been walking toward, GP/OCR-54575, according to Will's map. It's regenerated naturally after being high-graded, logged for only its best trees, a half century earlier. Sweating, I shake the dead fir needles off my hat and collar and step inside the stand. It's cool and the light sifts through the tree canopy overhead. Douglas-fir, bigleaf maple, and alder, some over a hundred feet high, make the stand shady. But sometimes, light floats to the ground and dances across the forest's floor like laser lights twirling in a disco. The rain of fir needles in the planted clearcut only a few hundred yards away is gone. All the trees are about a foot and a half around, but they are only half the size of the stumps they've replaced. Will tries to measure one of the trees with his diameter tape, but the tape is too short. He estimates its diameter using the end of Al's height pole.

Oxalis, sword fern, patches of bracken, downed limbs, and decaying logs cover the ground. I walk carefully around a patch of devil's club, a plant never found in clearcuts. It grows shoulder high, has large mealy leaves, and its stems are covered with fishhook-like spines that can break off in you. Northwest shamans prescribe the root of devil's club to cure an upset stomach and cancer.

Al figures this second-growth stand is about sixty years old and that the trees will be clearcut and trucked to the mill in another year or two. He's surprised they've stood so long. The trees in the planted clearcut we measured earlier this morning might only last a couple more decades, until they are thirty-five years old, before being harvested. "Why so young?" I wonder out loud. Foresters know that Douglas-fir are just beginning their most rapid growth when they're fifty to sixty years of age. It's longer times between tree harvests, not shorter ones, that increase timber yields and make the strongest wood.

When I moved to California in 1972, the rotation time between clearcuts in western Oregon was about a hundred and fifty years. By the time I moved back in 1984, foresters said that seventy years was more reasonable. I figured that an adjustment to the biological calculations of how trees grow in the region had been made, probably based on measurements like the ones I took with Will and Al in the planted clearcut earlier today. Now I know that most companies harvest every thirty-five to forty years, but the reason isn't because trees grow better, or the wood is higher quality, or that more wood is produced. Then why the short rotation, I wonder.

"Nothin' worse than a hungry mill," Will says. I catch the bit of sarcasm in his voice. "The company's gotta keep that big mill runnin' somehow."

"Yep," Al chimes in. "The rotation is about growing money, not trees. Our accountants hate having logs standing upright for too long—costs too much money, they say." His eyes twinkle at his own insight.

"Huh?" I say, stumped by both explanations but willing to play along with the notion.

"Look," Al chortles. "It's like Will says. Keeping that mill going is what's most important for the company, so our accountants have come up with a bunch of mumbo jumbo to explain why short rotations are best. You should hear 'em talking about interest rates and lost investment opportunities. They sound like a coven of witchdoctors preaching some kind of weird voodoo medicine, if you ask me." Pah, he spit on the ground. "Hell, we all know this land can grow three or four times more wood when the trees are seventy years old than if they're cut at half that age. Will and I measure these goddamn trees someplace in these mountains every day. We know."

"Why do it, then? Why cut so early?" I ask. Now I am puzzled.

"Use your head," Will retorts, miffed at how unaware I seem. "It's our job; we do what the company says."

"Won't the company run out of lumber as the rotation shortens and log yield goes down?" I ask, taking my chances with another dumb question.

"Naw," Al expounds, now really into is own argument. "Look around. There are plenty of trees out here in these woods. There are plantations everywhere. Besides, if the time comes when we can't grow enough timber on our own, we'll take over some other company, buy up their inventory. That's why Weyco took over Willamette last year. Hell, you can always buy lumber, always!"

I sort out what I've just heard as the three of us walk silently along the dirt road back to the truck. It's the mill that determines the interval between timber harvests, not forest biology. The fact that production goes down as rotation time shortens doesn't matter as long as logs keep rolling to the mill. The supply of trees is endless, according to these guys. True, I guess, but what Oregon loses is its forests. Will and Al didn't say anything about that.

The irony is that it's possible to have both timber production and a seamless forest without any clearcuts or short rotations. The answer is still standing in the second-growth stand we just left. Instead of clearcutting, foresters could harvest the trees selectively, thin the forest, taking some logs to the mill while leaving more space for the remaining ones to grow bigger.

"Why don't foresters harvest trees selectively and thin their stands?" I mutter under my breath while stashing the hardhat and surveyor's vest in the back of the truck. It keeps forests around, they're not cut down perpetually like with clearcuts—and like pruning my vineyard, thinning keeps the forest vigorous. It seems so clear, so straightforward. Selective logging means the timbermen might have to wait for the trees to grow bigger, I say, answering my own question.

Yep. Nothin' worse than a hungry mill.

Wood Left Behind

From my vineyard in the Willamette Valley, I see smoke clouds over the Cascades every summer. I counted six fires last August and thirteen the summer before. Gray smoke billows up, high above the mountain ridges more than a hundred miles away. The mushroom-shaped plumes stay for days, growing as each inferno gains its ground—until the weather cools or the backfires lit by firefighters finally join and the flames burn themselves out. Then a pall shrouds the hills as the smoke merges with the haze of our eastern horizon, turning the sky into burning coals at dawn and dusk.

In the early 1950s, when I was still a boy growing up on Washington's eastside, my Granddad climbed with me up the sagebrush-covered basalt butte behind Dad's farm. He held me high on his shoulder and pointed south into a brilliant scarlet evening. "That's Oregon burning," he said. It was the Tillamook Burn. The crimson sky lasted all summer, and reappeared the summer after that. I wondered then why Oregon burned every year.

It was four years ago that I drove up Bald Peter's Ridge with Jonny Dalton, when the fire there was still smoldering. Now, I don't wonder why our forests burn somewhere each year.

"We'll need these today," Jonny says, holding out the yellow long-sleeved pullover shirt and khaki pants—standard fire gear, identical to the clothes he already has on. "The fire moved last night and it's dropped to the ground, but the crews are still mopping up Bald

Peter's Ridge. If there's another flare-up this afternoon, we could be drafted to help out."

I change hurriedly in the men's room of the Forestry Building where Jonny works, pulling on the bulky pants, then lacing up my high-topped leather boots. The rough cotton shirt scratches uncomfortably as I pull it over my head and down across my bare chest and back. I strap a canteen and small first aid kit to my belt and fit the orange hardhat that I'll wear the rest of the day tight to my head. At the truck, Jonny stows the rest of our gear in the bed of the half-ton diesel that we'll take up the backside of Jackson Butte. "We probably won't need any of this stuff," he shouts over the whine of mill saws a few hundred feet away. "The fire crews are already mopping up the area we'll be in, but you never know when a fire will jump." There's a hint of anxiety in his voice as he loads the back of the half-ton with the axes, shovels, a chainsaw, and two fully stuffed backpacks of survival gear.

Jonathon Dalton is one of the foresters for this logging company. Its main office is in Portland, but the mill and forest are on the eastside, east of the Cascades. This trip was arranged weeks ago. We'd met several years earlier at a conference about wildfires and fuel breaks in California. He'd worked for the Forest Service then, but had since relocated to Oregon. We'd reconnected. Would I be coming out to Bend or Madras sometime this summer, he'd asked over the phone. He offered to show me some tree stands that he'd had thinned out with chainsaws—fireproofing, he'd called it. I jumped at the chance, wanting to see firsthand how his thinning operations worked, how they looked on the ground. All that was before the fire.

We head west into the foothills of Jackson Butte, first along the two-lane highway past Jump-off-Joe, then up a long gravel road, soon leaving the sagebrush scrub and juniper rangeland that surrounds the mill that Jonny works for now. We pass through miles of open ponderosa pine. Its understory of waist-high sage and bitterbrush are so sparse that you could drive a jeep through it. Tufts of bunchgrass and fescue stick through the dry lawn of this season's cheatgrass on the forest's floor.

Higher up the mountain, the pines give way and mix with other conifers—Douglas-fir, white fir, larch, lodgepole and white pines—but the ponderosa are usually biggest. Underneath, if the trees aren't too dense, are snowbush and whitethorn. These shrubs have bacteria and fungi attached to their roots that use nitrogen in the air. It's a partnership where both the microbes and shrubs benefit from the shared nitrogen. When the shrubs die and decompose or if they're burned, the released nitrogen improves the fertility of the soil, helping the trees grow. But they also compete with the bigger trees for water and can fuel wildfires when they grow too close together, like they do now. Then, when a fire starts, the flames climb through the entwined foliage of the brush and small trees into the crowns of the bigger pines, leaving a barren hillside of blackened naked poles in their place.

Centuries ago, the Indians here set low-burning fires in these forests to keep brush and saplings out, make hunting easier, and to keep their huckleberry fields open and productive. I remember riding with Granddad and Grandma along bumpy, dirty roads like this to pick huckleberries in Washington. The "crop in the mountains," they'd called them. The Indian fires and those started naturally by lightning crept through the open-grown pine and fir forests then, burning downed limbs, blown-over and injured trees, and most seedlings and saplings that grew up in the understory—thinning the forest. It seems to me that thinning forests, whether by fire like centuries ago or chainsaws now, is a lot like pruning orchards. The bigger trees, the good wood that survives the fire or chainsaws, always grow better and the whitethorn, snowbush, and huckleberry sprout again, rejuvenated by the disturbance.

Around the turn of the last century a series of decades-long droughts, combined with normal lightning strikes and intentional fires set by early settlers to clear the land, burned up most of the forest in this area. That catastrophe and ones like it all through the western U.S. caused a nationwide prohibition of fires in national forests, enforced by the then newly formed Forest Service. But the century-long policy hasn't worked well. The forest here grew back. But without the repeated low-burning fires of the past, it's now too thick; its understory is too dense. This stresses all the trees and makes even more fuel. So when lightning strikes now, after decades of fuel accumulation and logging practices that leave tons of dry slash on the ground, massive wildfires are set off someplace in the Cascades every year.

The road up Bald Peter's Ridge is dry and dirty. After more than an hour of jostling in the half-ton, we stop near the top of the ridge by an artificial pond used to refill water trucks during fires. Dust from the road swirls around us, leaving a deposit of grit over the truck and us as we get out. Two men are already here, also dressed in yellow shirts and khaki pants, leaning against the fender of their own half-ton. One man, Jim McAffrie, looks to be in his early sixties. He's lean with thinning sandy-gray hair. The other, Paul Lazarre, is younger, maybe forty-five years old or so, but already bald. Paul carries a hatchet on his belt and a hand lens that dangles around his neck by a leather shoestring. He's an entomologist from the Forest Service experiment station in La Grande. Jim is supervisor of the company's sawmill.

We're on a bench that wraps around the ridge. Several small ravines cut through the area, but the streams that made them are dry this time of year. The ground's surface is burnt to black dust, except where tree limbs and twigs have vaporized to white ash. Stubs of charcoal from burnt off shrubs stick haphazardly out of the blacken earth. Downed logs and stumps still smolder, even though the fire ran over the area days earlier. The smoke hangs like a dry fog in the air, along with the ash and dust we kick up with our boots. Surprisingly, most trees even in the densest stands are still standing, their needles a mix of green, gray, or rusty red. A trail of charred bark travels fifteen or twenty feet up every tree trunk. If a tree is small, a sapling or pole, all its foliage is dead, usually gone. The upright poles and barren limbs that remain resemble the black wire skeletons of broken umbrellas. On larger trees, the needles are rust-colored or scorched gray, but this damage usually doesn't go around the whole tree.

Paul hacks into the base of a large pine with his hatchet. Over half its needles are ruddy colored or gray, seared by the heat and smoke of the fire. He makes another cut in the same place, rifles through the chips and pulls a half-inch-long larva from the base of the tree bole. The grub is cream-colored with a brown solid head. "Turpentine beetle," Paul announces, the larva wiggling in his open palm as we crowd closer to see. The adult beetles hone in on injured trees like this one in only a couple of days after a fire. They burrow into the tree, winding around between the wood and bark, eating and leaving eggs. The eggs hatch and a new brood of beetles fly away or inhabit the same tree after a few weeks.

"After a light fire these bugs take out the small trees that the fire has already weakened. They thin the stand naturally," Jonny explains while Paul hacks into the base of another tree. "In dry forests like this one, beetles chew up the wood and speed the natural decay process along, recycling nutrients stored in the trees back to the soil." I remember that a colleague at the university, a microbiologist, told me once that a tree lying dead on the ground has more living cells in it from insects, microbes, and fungi than when it was upright and alive.

"It takes a whole bunch of these beetles to kill a tree this big," Paul continues. "Ambrosia beetles take longer to come into a stand, but it takes fewer of them to kill a tree. The beetles sense which trees are stressed most, even without a fire, by chemical pheromones that attract more and more of the insects. But a trunk needs to be ringed completely by their tunnels for them to kill the tree. If over half the tree's foliage is scorched, like this one is, it's weakened enough that the bugs can usually finish off the job."

"Uh-huh!" Jim grunts approvingly. He smiles as he talks. "I've got a crew of loggers all lined up for a salvage cut. They can be up here in the morning. The fire went over the ridge into the Forest Service yesterday. Nice stands over there, huh, Jon? It'd be nice if we could move the loggers in there after we get done taking down these burnt-out stands."

Damn hungry mill, I mutter silently. There are already too many trees going to that mill and others like it. Oversupply is why the price of logs is so low, the reason sawmills all over the region are going broke, shutting down, going out of business. Hell, this guy's not even sure that enough big trees will die here and he's got a salvage crew coming in the morning. So the Forest Service is supposed to let him salvage out their forest too? Why, to subsidize an oversupplied mill with still more trees, to keep it running? God, I even hate to think about doing such a stupid thing.

We continue uphill toward the ridge top. Along the way a mop-up crew, twenty young men and women, all wearing yellow shirts, khaki pants, and orange hardhats, march through. We see them up the ridge a hundred and fifty feet away emerging through the low smoke and haze. The black dust swirls knee-high as they walk side by side. All carry full backpacks, an axe or hatchet, and a shovel. They're two or three arm lengths apart, striding purposefully across the hill's contour in formation, heads down searching—step by step—for smoldering embers hidden in the still warm duff of the forest's floor. One stops and hacks out a pocket of smoke from a stump with her long-handled axe. The neckline of her shirt is wet, her face stained from the sweat and dirt. She swings a second time at the patch of smoke; then, satisfied the flame is out, she adjusts her pack and runs a few steps forward to catch the others and fill in the ranks.

Farther uphill, we pass through a clearcut that was probably planted with ponderosa pine a decade earlier. Clearcuts like this one fill in quickly with whitethorn or snowbush, after they're planted. Only foot-long burned-off stubs of the brush and the ruddy foliage or black umbrella-wire skeletons of the young pines are left. Fires usually burn along the ground in a brushfield, and here the pines were too small to escape the heat. In another week or two the brush will sprout from its roots, growing two feet by winter. Jon will replant the pines next spring. "Maybe enough fuel has been cleared away by the fire," he says hopefully, "so the new trees can grow big enough to escape the next one." It seems futile, this planting, then planting again, after every fire.

Next we come into one of the tree stands that Jonny thinned out a year earlier. The foliage on nearly every tree is seared and at each one's base there are neat little piles of a sawdust-like powder, frass, from the bark beetles already burrowing into them. The half-burned slash left over from the chainsaws still leans against charred upright stems of the scorched trees. Here, the dry slash carried the fire to the base of each tree, creating a ladder for the swirling flame to climb high into the trees' crowns, killing them all.

"The mill couldn't use the wood we slashed out of here," Jonny says. "I wanted to jackpot burn all this slash, hand-pile or yard it into the openings between the trees after the thinning. Get it away from the tree trunks, then burn it in the spring, but it was too expensive to do that. " Jim agrees.

"The mill couldn't even use it for chips or hog fuel?" Paul asks.

"Nah," says Jim. "The trucking costs eat you up. These little pecker poles can't even pay their way out of the woods."

"So I just left them where they fell," says Jon.

"Too bad," says Paul. "The bugs will kill any tree that's still alive now. Too much damage."

Still uphill and near a switchback on the dusty road that leads to the ridge's crest, we walk into a stand of wide-spaced ponderosa. Jonny explains that this is one of the first stands he thinned, over two years earlier. The ponderosas are far apart; their limbs don't touch. There are only a few of the other tree species left, a couple of big larch and a Douglas-fir. "If the thinned out trees weren't big enough to mill," Jonny says, "I cat-piled and burned them." He worried then, he said, that too many trees had been cut out. He'd wondered if the tractors used to pile the slash had caused too much soil compaction, or if they had damaged too many young seedlings and saplings to keep the stand growing. The fire had climbed twenty to thirty feet up the tree trunks, but less than half the crown of any tree was scorched. We wonder out loud if even that amount of fire damage is too severe; if the trees can rebound from the injury, escape the bugs. Paul thinks the trees are vigorous enough to survive the fire, even though he finds beetle frass at the base of nearly every trunk. I think so too.

"If in doubt, take 'em out," Jim says, "that's my motto."

"Maybe there's a better motto to use out here," I snap back, perhaps too quickly. "A log on the truck isn't the only good wood in a forest, you know. Maybe what you leave out here is more important than what you take."

"That's how I think about it too, especially for stands like this and the last one we were at." It's Jonny, chiming in unexpectedly. "I wish I'd managed the slash differently back in the other stand, spread it out or piled it away from the trees. I think it would look more like this one, resisting the fire and bugs better." I'm pleased that my friend and I agree.

But Jim wrinkles up his face in a smile that turns into a squint of disbelief at the way the conversation has turned. He mutters something about meeting a bottom line, but I don't hear it all. I'm already hiking back down the ridge.

It's September three years later, and in the heat of summer. The fires this year are farther south, nearer the California border. The news reports say they are the biggest in Oregon's history, bigger than even the Tillamook Burn that Granddad pointed out to me decades ago. Jonny and I are back on Bald Peter's Ridge. We're standing under the lookout tower on top of its crest. He's back with the Forest Service. He quit his company job a few months after the last fire here. A good divorce, he jokes now. Irreconcilable differences between Jim and me, he says with a wily, knowing grin referring to his former boss and the forest management philosophy he represents.

I look up tentatively at the narrow seventy-five foot ladder that leads into the trapdoor of the lookout station. The station's been vacant for years. The Forest Service canvasses the area for fires better with satellites and airplanes now. The persistent wind tugs at my shirtsleeves as I climb up behind Jonny. At the top, clinging to the ladder by one arm, he pulls a key from the shirt pocket of his uniform, shoves it into the padlock, and pushes the trapdoor open. It slams hard against the inside floor. We roll in, out of the wind. Inside are an oversized map desk, double bunk, and braces for a spotting scope mounted in front of each massive window. Jackson Butte stands out. It resembles a gigantic chocolate sundae with glaciers and barren rock avalanche shoots running down its sides. I'm not sure if the station is swaying with the wind or if it only appears to because of the tree movement outside. I feel woozy, a little seasick, but above the treetops I can see for hundreds of miles—even far to the south where several mushroom-shaped clouds have formed.

Below, remarkably, I see forests. I'd expected to see only a charred desolate hillside, the stark remains of dead trees. Instead, I see a mosaic of stands, some dead but most alive with trees of all sizes growing in them, and brushfields. Where the fire crowned the stands are gone, replaced only by dead tree skeletons and now the green understory of shrub resprouts, white-topped beargrass, and planted ponderosa. In other stands where the fire fell to the ground, ruddy colored canopies of dying trees stick out among the abundant green foliage of the fire's survivors.

"Looks like the fire and bugs are doing their work over here, thinning out the stand," I say, pointing to the west. "I thought it would look a lot worse." But Jonny's not listening. He's peering intently though a pair of binocs at the stands almost directly below that he'd thinned out several years earlier.

"Huh? Oh yeah," Jonny replies. "That's Forest Service land you're looking at. We can't clearcut because of the regulations in the Northwest Forest Plan. We can thin, but getting a logging contract through to do even that is unbelievably hard. The new law that Bush pushed through Congress last year makes thinning a little more possible, but the operation has to pay its own way. That means setting up a logging contract with someone like Jim MacAffie at the mill. Those contractors only bid if there is a high enough proportion of big trees in the stand that can be taken.

"Them pecker poles don't pay their way out of the wood you know," he says, mimicking his former boss's words from a few years earlier. "Forests like this should be managed using a different motto than Jim uses. How about, *leave the best, take the rest*? Instead of *if in doubt, take 'em out*, like Jim wants to do." I like the sound of Jonny's new motto.

"Damn it. I still think stand thinning makes sense"; it's still Jonny talking. "We know what trees to leave and learned to be careful with the slash. Thinning rejuvenates the trees left behind, like what you do by pruning that vineyard of yours, but not if the trees left are too little and suppressed to grow. Big trees need to be left standing, not logged out to pay for the thinning operation. Thinning isn't cheap or simple, I know that, but it is a lot cheaper and safer for people than wildfires." Then he adds almost as an afterthought, "Guess I still don't trust the whims of fires and bugs enough to completely turn the forest over to them. I want to be in on the action."

I look down at the stands below too. The whole area, even the stand he'd thinned and we thought resisted the fire, has been cut down and trucked to the sawmill. "I can't see how salvage logging or the current thinning law helps the forest out," Jonny continues. "It's the definition of 'dead and dying trees' that gets my goat. Guys like Jim McAffie and even one of my Forest Service supervisors have a different notion about what a dying tree is than me."

"A tree on the truck is a dead one, for sure," I say. "So, you think the new law is just a another excuse for the timber industry to get more loggers into the woods, the public forests, and to feed those tree hungry mills?"

"Yeah, I do," Jonny responds, "but whether the law is good or bad depends on how the logging contracts get written; how many big 'dying' trees get put on the trucks. A lot of the trees in these stands below us that were cut and taken to the sawmill probably should still be standing, alive on their stumps. It takes a lot of green trees to make salvage logging break even."

"It does in a thinning job too," I say. "So Jonny, is it the whims of nature or the motivations of timbermen that you don't trust?"

It's silent for a long time. I wonder if he's heard me above the wailing wind outside the tower. "We're still trying to make up for a lot of past mistakes out here," he finally says. "But you know, in a generation or two we won't be remembered for the profits made off this land. Our legacy is the wood we leave behind."

Starlings

While I was pruning yesterday near the little marsh on the corner of the farm where grapes won't grow, a winter flock of starlings flew into my vineyard. A thousand starlings suddenly lifted off from my neighbor's pasture a quarter mile away, causing the air to swirl as the black cloud of birds settled on to the trellises around me. Surrounded, I gazed silently at the birds for a second or two. Then as quickly as they landed, the starlings took off again, but this time they split into two clouds. Every bird had its place in the flock as each cloud circled overhead and coalesced again. I looked for a leader, some charismatic starling who the others follow, taking its lead to fly, turn, split the flock in two, or land. But there was no such bird, no chirp or sign to maneuver this way or that. Rather, each starling responded to the ones nearest to it as they flew in unison, one flock, a society of individuals. The flock circled low enough for me to see flecks of purple, green, and gray on the breast feathers of the closest birds, then it glided into the vineyard and I was surrounded by the clatter of a thousand voices.

Most of the year starlings live as pairs or in small flocks. They build nests each spring beneath the raised tiles along the edges of my roof and hatch their squawking fledglings there. They gather in my neighbor's orchard, roosting in the volunteer cherry saplings along the fence that separates our two farms. These birds sneak into the edges of the vineyard and steal the ripening fruit until only the picked-over skeletons of grape clusters are left by harvest time. During the last few weeks before harvest I scare them away with electronic sounds of starlings in distress or the calls of their predators that blast at one-minute intervals from four loudspeakers positioned seventy-five feet apart around the center of the vineyard. The last week before harvest is always the worst. Last year I plugged a radio into two hundred feet of extension cord and hauled it to the middle of the vineyard. I tuned it to Rush Limbaugh's talk show and turned the volume to as loud as it would go. Angry Republican voices shouting at the top of their lungs, along with the predator and distress calls, were enough to scare away the starlings. It also frightened almost all other animal life from the vineyard, including me, until the fruit was finally ready to pick in early October.

Why do people behave in a group in ways they won't do as individuals? For example, people would never crap in their neighbor's well, but the City of Portland dumps raw sewage into the Willamette River every winter. Most of us would not purposefully kill another person, yet our armies rush to war so readily. Why are sawmills shutting down and why are loggers out of work all over the Northwest? Why have some salmon species almost become extinct and why have our forests been burning down for the last seventy-five years? Why are we plagued by continually rotating clearcuts across our mountains? Ecologists say these are cumulative effects, the accumulated impacts of how timber is grown, how forests have been managed for the last three decades. I call them a tyranny of small decisions; the result of too many easy, simple, separate decisions that amounts to the same mismanagement. It takes a collective state of mind to cut down an entire ancient forest.

Forests, to some people, are a way to achieve corporate revenue. The volume of readily usable timber is their measure of value and their medium of exchange is a log on the stump, not any ecological, recreational, aesthetic, or spiritual value that forests also provide society. It's not that I'm against making money; I'd like to do it myself. Nor am I against using wood. Lynn and I live in a wood house. It's a home built for the weather of western Oregon. It's framed with Douglas-fir and covered by cedar siding to repel the rain. Decks of Douglas-fir surround it, and inside hardwood floors, living room ceiling, stairway, and moldings around our windows and doors are made of oak. Our cabinets are constructed of pine from the Cascades. It's the preoccupation of foresters with stumpage, board feet, and the notion of production for its own sake that ruffles my tail feathers.

Forestry to industrial foresters is equivalent to farming. They say to me, as many farmers have about their farms, that forest environments need to be controlled, manipulated, in order to grow more wood and maximize commercial investments in land, equipment, and people. Foresters, they say, improve nature by shortening the time it takes to grow and harvest a crop of trees. Nature, left to itself, regenerates trees much too slowly through the processes of natural tree generation and plant community succession to be of use to commercial forestry. This way of thinking when applied to growing food has not served farmers well for the last hundred years. It has resulted in high productivity but also fewer farms, displaced farmers, simplified ecosystems, eroded soil, and lower prices. How can the agricultural approach to production be expected to perform better for foresters than it has for the profession that invented it?

The agricultural way of thinking also affects how timbermen view old-growth forests. In their zeal to maximize production, they see old-growth forests as being wasteful because they are filled with "over-mature" trees. Logging is necessary, they say, to salvage this decadent timber to supply society's need for lumber and paper, and to also reclaim the land for more productive, fast-growing trees. But these fast-growing trees are the least valuable commodities that our forests produce. If Northwest forests are known worldwide for the high strength and quality of their mature timber, why are industrial foresters intent on growing more and more of the lowest-quality wood product possible?

I would like to believe that there is a grand conspiracy among foresters to maximize timber and every dollar invested in any industrial or public forest, that some misguided corporate executive officer has conjured up and directs all this poor management. But there isn't a conspiracy. Foresters, I think, are like the winter flock of starlings that flew into my vineyard yesterday. They maneuver in unison, manipulating the forest environments they manage just as each starling senses the movements of the flock by the motion of the other birds around it.

I have been in the societal debate between foresters and environmentalists over how timber is grown for every day of my career, as a scientist, teacher, and farmer. I am weary of it. Foresters and ecologists say a solution to our debate is too complex to be achieved, but complexity, it seems to me, is how sociologists describe the frustrating process of getting people to do what they don't want to do. Aldo Leopold, the great natural historian and conservationist of the twentieth century, foresaw this problem in 1938 and asked, "*Need we always await the willy-nilly pressure of wrecked resources before professional cooperation begins*?" Foresters know the economics of timber production. They are also taught the agricultural model of ecosystem simplicity and productivity, while learning about the value of diversity of natural forests. So while foresters and their teachers fly around in circles looking for simple solutions to fix our wrecked forests, the tyranny of small decisions continues and the consequences of how forests are grown accumulate year by year across this land.

The rules of a flock are simple: avoid obstacles, give enough room to your neighbors, get along. I spoke at a conference in Portland last winter about the diversity of perspectives of our natural resource debate, and the environmental and social consequences of our inability to resolve it. The audience, each member clothed in a suit of black or shades of gray and brown, were perched on rows of chairs, all equidistant from one another.

Foresters, I told them, all learn what is good for the land, for the creatures that inhabit forests they manage, and for the corporate pocketbook. It is the compromises necessary to have all these benefits that are said to be too complex, the obstacles to resolution. There were no caws or cackles during the speech; no one asked me a question at its conclusion although prompted by the moderator. But when it was time for the session break, they all swirled to the back of the auditorium for coffee or tea. They stood in small groups of threes and fives, and once again I was surrounded by the clatter of a thousand voices.

Part Three

Renewal

A package came in the mail from my sister Annie last Christmas. It was wrapped in plain brown paper and tied tight with string. Inside were two miniature chairs and a tiny table, all no more than a few inches tall. They were made from black walnut twigs, some of last spring's pruning from the trees we'd planted on the homeplace almost a decade earlier. She'd carefully nailed and glued the sticks together, then tied a note to the leg of one of the chairs.

It read, "I thought you should have the first furniture made from our walnut trees." The little chairs and table sit on the windowsill of our kitchen at Kla-kla-nee.

Kla-kla-nee

The Kalapuya tell a story about the Three Sisters, the three ice-covered volcanoes that rest together in Oregon's central Cascades. They say that when all the world was forming, three giants, sisters, were making their way north to rendezvous with the Great Spirit. They were to gather, along with other giants, to consider the merits of becoming mountains. The three from the south had the farthest to go, so they hurried but along the way the smallest sister got tired and straggled behind. The other two sat to rest and wait for her. The smallest, seeing her older sisters resting, sat too. They never moved again. The Great Spirit tired of waiting and turned them all into mountains where they sat. Lynn and I have lived in many places of the West, from Washington's high desert to the golden floodplain of the Sacramento River. Now the emerald valley of the Willamette is our home. We call our small farm here Kla-kla-nee, the Kalapuya name for the Three Sisters, because on clear days in the Willamette Valley we can see these three snowcapped travelers from our home. They are rooted, in place, like we are.

Our farm has four acres of vineyard and a small orchard with two varieties of pear, three types of apple, a plum, a prune, three peaches, and a fig tree. There's a patch of blueberries and several beds for vegetables that we plant anew each year, and Lynn scattered wildflower seed several years ago that now grows all around. Sedges seeded into the half-acre that's too wet for vineyard, so I've planted willows and a dead cottonwood snag there for the bluebirds, wrens, and blackbirds to roost in. A yellow rose, transplanted from our other farm in Tieton, trails along the fences that surround this place. Our two sons and daughter have grown up, moved away. Steve and Matt live in Vancouver, Washington, now and Susan moved to Thousand Oaks, California, a few years ago, but this farm is theirs too. It's the place they come back to.

It took us three weekends during a soggy January in 1991 to dig up our new grape plants from the mud of a tiny nursery near Donald. Susan had just returned home to reenter college, so she helped Lynn and me shovel out the young vines and separate their roots from the mud with our fingers. We spread the cuttings on the ground for the pounding rain to wash clean. Then I tied them into bundles of twenty-five, packing them in moist sawdust until spring.

One morning we gave up our digging early. Cold and wet from a particularly hard winter shower, we decided on an early lunch at the Donald café, a few miles down the road. Most country towns have a place like this where farmers wait out storms, drink coffee, and talk about their kids, tractors, or the prices of crops. We made our way to the back of the restaurant, shed our soaking raincoats, and asked the waitress for hot coffee. Two farmers sat nearby, both wearing dirt-stained baseball caps; one cap had *Willamette Seed* written across its front, the other said *Farmers Credit Union*. I tried eavesdropping on the pair while Lynn and Susan went to the restroom to dry their dripping hair. It made me feel good, finding the café, listening in. Even out of easy earshot, I sensed what the farmers were saying by their language, its gestures, expressions, and tones. I'd grown up on a farm, been a farmer for a little while, but moved away to California and became a professor. I'd started to take farming for granted, lost touch with the awe of growing things. But I came back. That's why I wanted to plant this vineyard, to get back to the ground and begin growing once more.

I'd already laid out the rows for the new vineyard the summer before by plowing strips of bare soil ten feet apart, north to south, through the old fescue sod of the abandoned hayfield we'd bought. Planting took all April—Lynn, Steve, Susan, Matt, and soon-to-be son-in-law Tim—all planted whenever we could. I showed them how, and we worked in pairs; one digging while the other planted each rooted stick into a freshly dug foot-deep hole. One late afternoon, Tim, who is from L.A., asked if I'd show him how to grow corn sometime. "I don't think I've ever grown anything, and then eaten it," he explained. "Sure, it's not hard," I responded casually, but wondered how many others might say something like that; not many people grow their own food anymore.

There were only fifty or sixty plants left on what turned out to be the last day of planting. I was working alone; it was late and even my hair felt tired. Steve, who was a senior at OSU then, found me at the farthest corner of the field. He'd skipped his last class and driven out to help. "If you plant, I'll dig," he offered. We finished planting that evening in the dark.

The vineyard is grown up now. I spray it in the summer for mildew, and prune it every winter weekend. I sometimes park my John Deere in the vineyard while I'm pruning; it gives my grandsons—Tyler, Timmy, Kyle, and Nick—something to climb on whenever they visit me out there. It's only a three-cylinder diesel, half the weight of the old 8N Ford I used to have, but it's more maneuverable and has the same amount of power. A farmer I know in California has a tractor so big it's more like a place than a thing. Tractors work best when they fit the size of a farm. Too much power, I think, is not a good thing—for the land or people.

Every October the vineyard produces two tons of Pinot Noir and six of Pinot Gris, depending on the weather. I hire a contractor who brings in a crew to do the harvesting. We sell the crop to the little winery in Bellfountain, not far away.

I usually work in the vineyard alone. Lynn says I enjoy the company. She's right. Besides grapevines, there's cat's-ear, groundsel, Queen Anne's lace, cow parsnip, tansy, spearmint, ryegrass, and tall fescue—all left over from the old hayfield—that grow in my vineyard. There's also sheep's fescue that I sowed between the grape rows to keep down the dust in summer and hold in the soil and water during winter. Lately I've noticed sorrel, blackberry, California poppies, air-grass, and a little pretty blue fiddleneck coming in. Meadow mice and garter snakes, blue darters, pocket gophers, cottontails, bats that I can only find at night, finches, swallows, and bluebirds live in my vineyard. Sometimes raccoons and little red foxes scurry through. A family of skunks and a packrat built their nests in the blackberry brambles along the back fence. Each morning in the spring, a pair of redtails graze over the vines, hunting mice and gophers between the rows. Turkey vultures circle overhead on hot summer days when the convection currents are right to take them high into the wind. They leave for Mexico or Argentina before harvest-time.

Grasshoppers, leafhoppers, thrips, mites, and spiders climb among my vines. Praying mantises hunt for them, camouflaged in green when the foliage is lush but changed to tan in the dry summer grass between the grape rows. There are red earthworms, blue earthworms, nematodes, *Botrytis*, bacteria, and algae in the soil. All these things exist, live, and thrive in my vineyard. They belong there, like me, but I wonder what to make of all this company.

Blackberry and grape plants compete for water, but less water, especially in late summer, makes better wine. In winter, aphids, mealy bugs, and their insect predators hide in the blackberry foliage after the grape leaves drop off. Though low in number, the predators keep the other insects in check there, so a pest epidemic never starts when the weather warms and they all move back into the grapevines. I used to trap gophers. I probably caught fifty of them, but stopped when finally I realized they never eat grape roots. They turn the soil over and dine on cat's-ear and clover. Once I dug into a gopher tunnel and found a cache of cat's-ear roots, yarded and stacked like cordwood, fermenting like silage in a silo for winter. The deep roots of Queen Anne's lace help the winter rains penetrate deep; fescues keep the soil from eroding into the river and mud off my tractor's tires. When these plants die and decay, they give worms and bacteria something to chew on and enrich the soil. I spray against powdery mildew several times each summer; it rots the young fruit in their clusters if I don't. Spearmint and sorrel stay in the wetland where grapes won't grow.

Before settlers arrived here, every year the Kalapuya burned this hillside where my vineyard is now. Elk, white-tailed deer, and woodland bison grazed it. My neighbor says there's still a buffalo wallow where the grass barely grows just across the road from his cherry orchard. Only a few thousand acres of that native savanna remain, but I'm only trying to mimic, not recreate something natural out of my vineyard.

Most growers now try to increase their yields by making their farms simple, less diverse, to have fewer species. These farmers and the scientists who advise them say that too much diversity of plants and animals causes complications, complexity, and competition. Manage simply, they say, but end up mining their land, then trying to subsidize the loss with chemicals and money. A few farmers, like me, keep more plants and animals around. We like the diversity and think it contains diseases, prevents insect epidemics, improves the soil by recycling its fertility, and maintains productivity. But I really don't know if either notion is true. Too many explanations make nature plain.

I don't want to construct anything deeper about the organisms that live in my vineyard, or their ways. I only know they belong there, like me.

It's been over a dozen years since we planted the vineyard at Kla-kla-nee. This is the place our kids and grandsons come back to. When Tyler, our oldest grandson, was five years old, he found a finch's nest secluded in one of the vines. Finches build compact, round little nests of dry grass among the summer canes that surround and hide it. I wrapped a red plastic surveyor's flag around the top trellis wire to mark the nest, and Tyler checked the tiny white eggs daily. He watched the eggs hatch and saw the tiny birds fledge and finally fly away. Now I mark all the nests in the vineyard; it's easier to see them that way and to shut down the sprayer as I spray the vines for mildew. The next spring, Lynn and I nailed six birdhouses to some of the trellis poles to attract bluebirds and swallows, which they did.

Three years ago when Tyler was nine years old, he and I pulled a knotty cottonwood snag from my neighbor's burn pile with the John Deere and planted it upright in a deep hole to attract more songbirds. Tyler and his cousins, Timmy, Kyle and Nick, call the snag and little wetland around it the bird sanctuary. I helped them cut willow twigs and plant them along the ditch they named "Little Creek," so the blackbirds can roost in them. The patch of blackberry along the fence is called "Briar Patch" because rabbits and foxes hide in it, and the big Douglas-fir that grows out of the middle of Charlie Fischer's cherry orchard is known as "Hawk-tree." A red tail roosts there, eyeing the orchard and vineyard for unwary mice and gophers.

Two summers ago, a week before harvest, I moved my 160-gallon plastic tote bins to the edge of the vineyard and set them upright to form a maze. Steve's three boys pushed the heavy bins around until they made a fort with two entrances and several rooms. They foraged the garden, orchard, and blueberry patch, and raided Lynn's kitchen for provisions.

I contract with my friend Luis to harvest the grapes. It's the only labor on the farm that I don't do myself. Last year he brought in sixteen pickers, and the job was done by evening, except for a couple of short rows that I intentionally keep out for our friends to pick. They come over an hour or two before the sun goes down to harvest the last of the crop and celebrate the end of another growing season, the start of a new vintage.

Even after the picking is done there's plenty to do; bring in the loaded bins from the vineyard with the John Deere, load the truck and trailer, tie down the full bins on it. Take them to the winery in the morning. A couple of years ago, it was already dusk by the time picking was over. Our friends were accumulating at the house for a potluck dinner and to taste last year's wine. Later that evening, we would stomp the grapes they'd picked in a washtub to start another year's vintage. I call the wine we make ourselves our mystery wine because it's uncertain from year to year how well it will turn out. But no matter, our guests always like it best. Lynn, Susan, and Tyler would be there with them, I knew, celebrating; so would Teresa and our other three grandsons. Someone was singing, playing a guitar at the house. I could see a fire in the chiminea flickering in the dark. The party had started without me, but that was expected. I worked my way in slowly from the vineyard with the lights of the overloaded tractor on. Steve and Matt were already strapping down the bins on to the nearly loaded trailer.

I backed the last bin of fruit on to our cement loading platform, then positioned it over the trailer. Looking backwards, I saw my two sons toss across the last strap as I lowered the bin. They ratcheted it down tight to the trailer. The job over, we ran our hands across the freshly picked fruit. It was fragrant, sticky, and full of the feeling of success. Another harvest done, we walked through the vineyard and orchard to the house to join our family and friends at the party. It would go on late into the autumn night.

High Ground

I gauge the intensity of our winter storms at Kla-kla-nee by the level of Granger Lake, which isn't a lake at all, only a low-lying swale between an unnamed creek and the high ground where Lynn and I built our house. Elton Logsdon, the wheat farmer who owns the low land beside Granger Road, plants his field each year but our seasonal floods control his crop, not him. Usually Granger Lake barely fills, but when it does migrating dusky geese hold over there, grazing the wheat along its gentle slopes, indifferent to the falling rain. They clatter into the air every morning just after dawn and circle noisily around the vineyard where I spend most days of my winter weekends, pruning vines between storms.

Most of our storms come in from the west, cold and straight off the north Pacific, rolling over the Coast Range into the valley. They hit the coastline in pulses of five or six in a row, each lasting two or three days. Then, as if worn out, they take a breather, resting for a week or two until another series starts up again. I admit it. I've gotten used to the winter weather here. Sometimes during a winter dry spell I even get anxious, wondering when the next set of freshets will sweep in off the coast to irrigate my dormant vineyard. I like watching raindrops splatter across the double-paned windows of our home, hearing them pound against the metal sheets of my barn's roof. I enjoy lying in bed with Lynn listening to the wind push uneven gusts of rainwater against our home, hear water rushing down our pitched tile roof and rattle through open gutters only to sink silently into the vineyard's deep soil. I can prune through a rainstorm if I want to, wrapped in cloudy green raingear—waterproof jacket and pants, sturdy rubber boots, a broad-brimmed felt hat pulled low over my forehead. It's steady work, like the rain.

But during January 1996 a storm blew in from the south that was different from the rest. Its warm wind followed on the heels of a long cold spell that dropped more than a foot of snow in the Cascades to the east. The rain came in nearly horizontal, pushing through the seams of my rain jacket, rolling down the back of my collar, soaking me from both inside and out. I gave up pruning early that morning and shoveled out all of last year's drainage ditches around the vineyard, clearing away the weeds so the rainwater could flow quickly to the creek beside Granger Lake. The wind was so warm, the rain so intense. "This storm could be a real downpour, a fence raiser, a gully washer," I said to myself as I walked back to the house after less than an hour, drenched from the driving rain and my own sweat.

Lynn and I watched through our living room window as the storm grew fiercer. A small branch broke loose from a birch tree on the other side of the driveway. It flew away, landing in the vineyard and tangling in the wire trellis. The house shuddered as the main front hit, blasting gust after gust of solid rain against its cedar sides. Inside, I adjusted the damper on the wood stove and laid another chunk of oak on the smoldering fire. Lynn found a worn quilt that that she's been intending to mend. Grandma sewed all of its pieces together, then gave it to her when Steve, our oldest son, was born. Now our grandsons choose her blanket to wrap up in on cold, damp days like this, perhaps feeling her legacy of love and warmth, even though they never knew her. That afternoon, Granger Lake started to rise and the duskies hunkered down to weather the storm; so did we.

I'm told that snow is such a common feature in the lives of the Inuit people, who live in the far north, that they have nearly fifty ways to describe it. Why then, here in the Willamette Valley, do we only have a few ways to say rain? We don't have words for our gentle drizzles that last for days, or the dark gray clouds that hang around heavy between squalls. We don't know how to say, "mist that condenses on your eyebrows until rivulets run down your face," or to describe gusty storms that splash raindrops hard into west-facing windows and collect as ever-present mud puddles. Maybe the Kalapuya, who lived here for centuries before us, have such words for the weather but I don't know them.

It's now the third day of steady rain. The lake has engulfed the entire wheat field and is edging into Charlie Fischer's cherry orchard on the slope below us. The water surrounds Logsdon's old farmhouse. It now stands isolated on its little knoll in the middle of a sea of brown water. The creek stretches more than a mile across the valley floor until it bumps against a distant row of cottonwood and ash trees that line the dike holding in the west shore of the Willamette River. On its other side and where the dike is lower, the Willamette floods over another ten miles of farmland. The unrelenting rain has melted the snowpack in the Cascades, which makes the rivers swell even more. People in Portland are shoring up their tidewall along the Columbia with plywood sheets and sandbags. The sewer in Oregon City has overflowed. A woman has drowned in the Luckiamute River near Independence, and a Corvallis friend rowed his drift boat down Third Street. Last night, the lights of Fischer's house reflected off the water creeping through his orchard. Lynn and I wonder about the geese on Granger Lake.

Most roads around us are flooded now. We're cut off on three sides. Only the narrow gravel road over Logsdon Ridge to our north still connects us with Highway 99. Not that we need to leave; we're safe here on this rise. Granger Lake is still a good twenty yards from the lowest corner of the vineyard, four hundred feet downhill from the house. There have been worse floods here in the valley. Orleans, a town that once thrived on the other side of the Willamette from Corvallis, flooded out twice between 1880 and 1890. After the second flood, the people there gave up, left a sign on the old church that now marks where the town had been, and moved to higher ground. And ten thousand years ago, the Missoula floods would have been most fearsome of all. A prehistoric glacial dam in what is now Montana broke loose sending a wall of water, rock, and ice hundreds of feet high down the Columbia Gorge. The water rushed into the valley, pushing up its streams, inundating all the ash swales, lowland meadows, and swamps.

When I built my barn a few years ago, I dug four five-foot-deep holes to cement the corner poles into. My shovel hit something solid near the bottom of the third hole. By the way my shovel cut around the object's edges, I could tell it had at least two squared-off corners. It made a hollow sound each time my shovel scraped across it. That sound and the shape of the thing in the hole made me think it might be a box. I didn't realize until later that my shovel's handle was cracked, which made the vibrating sound. I dug for nearly an hour in the mud on my belly with the shovel and then a trowel—convinced I'd found a lost strongbox or buried treasure chest—only to pull up a heavy rectangular hunk of granite that was about a foot square and flat.

Glacial erratics, chunks of granite like the one I dug up, are unusual finds in the Willamette Valley. I figure the one I found floated in from Montana, ending up on my farm after one of the Missoula floods. The ancient glacier it was imbedded in scraped its sides square, and the flood carried it here thousands of years ago when the ice dam broke loose. It stayed buried under four feet of silt, also deposited by the flood, until my shovel hit it. Now I use the rock to block my barn door closed so it won't bang against the wall when storms blow in. People lived in the valley during the Missoula floods, probably even on this little hill where Lynn and I live now. Perhaps they saw the gigantic wall of water and ice coming. Maybe they heard it first in the distance and ran for higher ground to the north or west. Perhaps.

When I planted the vineyard twelve years ago, I found two pieces of native basalt lying in different places on top of my recently tilled-up soil. Now they sit on the windowsill of my den. One of the pieces is shaped like a sitting bear with notches carved into both sides where the haunches are. My thumb and forefinger fit perfectly along its grooves on each side. The other piece is carved in the shape of a harbor seal and curved to run along an index finger to protect its tip. These are the tools of a patient people, a hide scraper and a miniature awl or the ancient version of a thimble—used perhaps to soften a buckskin seam. They make me think of a family hunkered down in the midst of a driving rain, sitting beside a fire barely outside their rawhide lean-to. These were people content to stitch a robe or mend a buckskin blanket while they waited out a storm. These artifacts make me wonder what the Kalapuya call a hard horizontal rain that blows in from the south and lasts for a week. They make me think, in the midst of this flood, that our small knoll has always been a safe place. A little high ground sitting above a flooding creek and an ephemeral lake where geese feed. A place to build a shelter, stoke up the fire, sew for awhile, wait out a storm.

Wedding Gifts

It's two in the morning and the reception of our oldest son's wedding is still going strong. I know first-hand how Teresa's father feels because my daughter, Susan, married only a month earlier. But now, Lynn and I are tired, worn out from the celebration, and my feet hurt from these stiff, black-polished shoes that came with this rented tux. We watch the couples slow dance to a Garth Brooks song—"Shameless." Steve and his bride sing the words to each other, swirling with the music. *I've never been in love like this,* they say, and sway with the lyrics of the song long after it ends.

There is a time, I think, when men identify more with the father of the bride than the groom at weddings. It's a turning point, one of life's transitions, a threshold. It happens when you recognize one day that your little girl will not stay that way forever; that someday you too will give your daughter away. I once asked an older friend and father of two beautiful married women about the inevitability of this feeling. "Best get used to it," he affirmed; "comes with the territory." The purpose of weddings, I think now, is not so much a celebration of two young people about to share their lives together as it is an initiation rite to prepare fathers for their daughters' leaving.

Susan met her future husband during the summer of 1990 in chemistry class. She had just returned from Michigan, a retreat of sorts, in time to enroll in summer school. She'd unexpectedly dropped out of college in California two years earlier and had temporarily returned home. I met her at the bus station; she looked drawn, troubled, sick. She soon withdrew inside herself, but after a few months agreed to move with Easly, our golden retriever, to Shaw Island in the Straits of Juan de Fuca. Worried, I'd arranged for her to help the Benedictine Sisters there work their organic dairy. She'd try it for a month or maybe two, she said, but stayed for a year. She arose early every day to milk the Jerseys by hand, helped birth their calves, delivered milk to the islanders, put up hay, and learned to fix farm machinery. She prayed with the nuns and, though living with two other girls, used the rest of her time to think, read, and sort herself out.

After that good year, she followed her two friends, who planned to join a convent there, to Michigan. Susan wasn't so sure, but packed her car anyway and, along with Easly, set off across the country. I came out a few weeks later to see her at the convent, the result of a timely collaboration between OSU and Michigan State University. Alma, the town she was living in, was only forty miles from MSU.

Susan met me at the airport. It was cold, winter was settling in, and ice was all around. Bundled up, we wandered around the MSU campus, visited the Vet School and got enrollment forms to the university. Susan and her friends were pleased with the boarding house where they lived across from the convent. She was enrolled in a local junior college and planned to take courses at MSU the following term. She showed me where the three of them slept and ate, and the tiny living room nook where she studied on a rickety card table. She'd taught Easly to heel; the nuns like obedience, she explained, and it helped in the city. But I sat in the Lansing airport the next day brooding, unsettled, silent. It hadn't felt right leaving her there. She seemed out of place, set aside, isolated. Reluctantly, I got on the plane to Oregon; there was nothing else to do.

I brought Easly home two months later. I was needed again at MSU and Susan called a week before the trip. "Easly isn't doing so well," she explained, near tears, stubbornly keeping them away. "My street's too busy and he's tied up all the time. He belongs in the country, where he can run. Will you fly him back when you come out?" she asked. After working at MSU most of the week, I drove to Alma for the weekend and to pick up Easly. It was early spring, and the nuns wanted to spruce up the convent's woodlot. A series of winter storms had left the property littered with downed limbs, branches, leaves, and paper from the busy nearby street. We enjoyed the work, Susan and I, along with half a dozen nuns from the convent. The crisp wind blew in our faces and the scent of burning leaves and dry wood from our bonfire filled the air. But Susan stiffened inside the convent. Her movements betrayed her frustration, her annoyance with the convent's hierarchies. She resented its pecking order. She irreverently referred to the retired bishop who also lived on the grounds as 'His Dibs.' I chuckled at that, but knew she was determined to carry on with her plan, whatever the hell it was. Easly's not the only one that belongs in the country, I thought, once again sitting in the Lansing airport. Silently, I contemplated ways to coax Susan into the travel container alongside her dog. Then I'd take them both home, that's for goddamn sure!

The next time I traveled to Michigan was on a return trip from Toronto. Alma was too close not to swing by, and I figured it might be my last trip east for quite a while. Susan greeted me at the airport, but before we were off the jetway she started jabbering. Excited, anxious, she jammered out, "Dad, I've been thinking a lot—about coming home." Did my heart just skip a beat? Did I hear right? We spent the next two days making sure she left Alma well, with grace, good feelings. We paid her bills, picnicked in the woodlot with some of the nuns, and built a stone altar in the convent's courtyard for 'His Dibs.' I left for the airport before Sunday mass ended. The Mother Superior of the convent slipped out too, wished us both safe journeys, and kissed me on the cheek. Susan left the next day and five days later, she was in Oregon and enrolled in summer-term chemistry.

A week before Susan's wedding, I awoke at dawn as usual, but lay tossing. "What's with you?" Lynn asked, awakened by my restlessness.

"Dads," I grumbled, "should do more than just give their daughters away and pay the bills." (My older friend told me once that the real reason God created proms was to prepare fathers for the sticker shock of the wedding. He was right.)

"What do you want, then?" she asked. Lynn had been heavily involved in the wedding, planning it with Susan; designing flower arrangements that she would make herself, sending announcements and finding people to cater, make cakes, and keep the guest lists.

"Tim gives her the rings," I mumbled. "Maybe I could do something special too, if I knew what. That's all. "

"Why don't you go with Susan to buy her dress this afternoon? She's narrowed it down to just a couple, and there's one she really likes. You could help her choose and buy it then," she suggested.

I rearranged my afternoon and met Susan at the Bridal Shop—a pretty little store filled with lacy things, silk evening dresses, and a white wall of bridal gowns. I fidgeted, feeling uneasy, clumsy, in the delicate little shop. Susan held up the dress she liked best, examined the beads that surrounded its neckline. Then she went into a back room to put it on. The dress rustled as she left the dressing room and stepped in front of a full-length mirror. She spun around twice. I hardly remember the gown. When did she grow up, become so beautiful? Bronze hair. Bronze skin. Brown eyes. The white material flowed around her when she walked, twisted divinely when she turned. When did she become a woman? The gown was perfect. She was perfect.

Tim's mother, Fong Su, and her husband, Eric, flew up from LA two days before the wedding. The ceremony would be in the courtyard at Hanson's Country Inn. It rained steadily the day of the rehearsal. Susan and Tim were late, so the rest of us fretted, talking over what we should do if it rained the next day too. We decided it would be best to move the wedding inside; but when Susan arrived the room quieted. She looked at me, puzzled.

"Susie," I said, choosing my words carefully, "we're wondering what you want to do if it rains tomorrow."

"Oh, it won't rain," she replied with youthful optimism. "Besides, Mrs. Hanson says the attendants could hold umbrellas over us if it does."

"That's right. It's happened plenty before," said Mrs. Hanson. "It looks kind of cute out by the gazebo with the maid-of-honor and best man holding umbrellas over the bride and groom."

Then, Susan said, "We could all just crowd into the Inn if it rains real hard, but it won't—right, Dad?" as if I somehow controlled the weather.

"Right, Sues," I responded, using her childhood nickname, and nodded with make-believe assurance. Lynn crossed her fingers.

Matt missed rehearsal even though he was a groomsman. His high school baseball team was playing that evening and he was needed to pitch relief. He'd meet us at the restaurant after the game, he said. Eric nervously started the toasts to Susan and Tim that evening. We were still mostly strangers. Then I welcomed Tim to our family, and Fong Su and Eric too. The evening went on. Steve, with Teresa sitting beside him, playfully warned Tim about Susan's stubborn streak, her flashy temper, her well-formed sense of rectitude. We were well into dinner when Matt showed up wearing his dirt-stained baseball uniform, hat on backward, cleats under one arm and his girlfriend, Carrie, firmly attached to the other.

As we finished dinner and waited for dessert, I looked at my family seated around the table and reflected about the future of each of my children—Susan and Tim's wedding the next day, Steve and Teresa's wedding next month, Matt's still a far-off possibility. I knew that each of them was making their decisions, pruning, keeping their own good wood, while letting go of the rest. Lynn and I had a hand in how they make their choices, about friends, professions, spouses. I wondered if they knew how much we believed in them.

It hadn't been an easy spring for Matt and me. He'd made the State of Oregon select-football team the fall before. So over spring break, he'd traveled with the team to England, playing exhibition games around the country. He had trouble being just a high school junior after London, and got a D in biology. We'd had hard tough words the night before; I had insisted that he honor his curfew and retake biology during the summer. We also talked about choices, commitment, effort, responsibility. It was our first discussion like that since his sixth-grade year. He listened conscientiously, head down, while I scolded. Finally, he choked out something like, "I won't let you down, Dad." Damn it! It wasn't me he was letting down. Couldn't he see that?

"Matthew, give us a toast." It was Susan calling to her younger brother from across the table. He'd just ordered a dessert for Carrie and dinner for himself. "Come on, Matt, give me a toast." She was having fun putting her brother on the spot. Self-consciously, he reached across the table, poured himself half a glass of wine and replied, "When this is gone, Sue, then I will." But with our dessert almost gone and his dinner started, Susan wasn't about to let her little brother off the hook. "Okay Matt, your wine's almost gone now. Give us a toast." Matt sighed and pushed himself laboriously up from the dinner table with both hands. Holding the still half-full wineglass, he said, "Sue, this is to you and Tim." He looked directly at his sister. The room quieted. The waitresses stopped clearing dishes. "I wish you both the best of life." It was little brother, speaking directly from his heart. He paused. "And when you have a kid, teach him. Teach him to play ball—I'll help with that. But mostly, teach him like Mom and Dad taught us. You know how, Sue." Then he sat down. Someone said, "Hear, hear"; maybe it was Eric. Susie and Lynn had tears in their eyes, so did Fong Su. As for me, there wasn't anything more that I wanted to hear.

Susan's wedding day started in a drizzle. Lynn and I went to Hanson's Inn early, taking in the flowers that she'd arranged the day before. Eric and Fong Su helped me set up chairs, arranging them in straight white rows. Lynn put the finishing touches on the wedding bouquets at the gazebo. Then, black clouds rumbled in, followed by sheets of sideways rain. The four of us ran to the Inn's barn for cover. "Hope those flowers survive," Lynn exclaimed, looking up at the clattering tin roof overhead. We all had our fingers crossed now.

The ceremony was short—so short that one of my sisters wondered if the newlywed's kiss took longer than the wedding. But the storm had broken and the sun shone through ivory-colored clouds that drifted in from the west. Steve and Matt looked elegant in their white tuxes as they seated their grandparents, Dad and Mom walked in together, Great Grandma, and then Lynn. Tim wore his Navy whites. All the bridesmaids walked slowly through the courtyard; then Susan and I, arm in arm. When we reached the gazebo, Susan turned and kissed me on the mouth. She hadn't done that since before she was a teenager. Lynn had kissed her Dad at our wedding twenty-five years earlier. I wondered if Susan knew.

"Thank you, Dad," she whispered, "for everything."

I muffed my only line, stumbling on "Her mother and I" after the minister asked "Who gives this woman to this man?" God! How hard can it be?

Safe Sites

It took nearly ten days to settle Susan and Tyler in Santa Barbara. Our daughter and grandson had been with us for over a year and a half, the consequence of a marriage left undernourished by the long absences that can accompany Navy life. Now they were moving away. It took me a day to load the twenty-six-foot U-haul, and there was an ugly lump of raw meat in my throat the whole time. It rained that afternoon too, a gentle Oregon drizzle, and I was glad for it. I wanted Susan to hear again how the rain settles across our vineyard, smell it fresh from the garden's new-tilled soil, remember the climate of her homeplace. Tyler and I negotiated. He finally agreed to leave his pedal John Deere with me, on the farm, next to my big tractor. Tractors belong on farms, we decided with misty eyes. There has to be something for you to come back for, I whispered, holding his little chest tight to mine. Our caravan left the next day, Susan and Lynn in the lead; Tyler and I followed in the U-haul, the big truck, he called it. It rained until we reached California.

There is a notion among ecologists of something called a safe site. It is a combination of the right genetics, environment, resources, and luck that assures success for dispersing offspring. Common groundsel, for example, is a small annual weed in the sunflower family that inhabits recently tilled soil. No more than a foot tall, it is fragile to look at, with slightly serrated leaves attached to wispy stems that sway constantly in the wind. It matures quickly and produces abundant seed from bright yellow flowers. Attached to its nearly inconspicuous seed is a delicate white parachute, an adaptation to enhance dispersal by wind. As the seed blows across the soil, the parachute snags onto plant debris, so the seed eventually settles into a small crease or crevice among the clods of soil, hidden. The parachute then absorbs water from the air and soil, encasing each seed in its own moist environment—a safe site. Safe sites, unfortunately, are elusive in nature. Some scientists even doubt they exist, saying the only criteria established for them are after the fact, when an emerging seedling eventually survives there.

In the plant kingdom and, I suspect, the animal kingdom as well, the safest place for offspring to grow is rarely under their parents. Parents, because they are large and obvious, sometimes create conditions unfavorable for their offspring, or they attract predators, parasites, or disease that can harm or kill the young, small, or unaware. Seedling shrubs and trees rarely grow well under their parents' shade, for example, and also run the risk of being eaten by birds or mice that use the parents' branches for cover. So for most plants, species advance as a front with progeny moving along from year to year, always ahead of their parents but in more or less the same location. The other successful strategy is to scatter seed widely, expecting that some will find a place, get lost in the crowd, and grow unnoticed, safe. The few that do survive establish new populations, new colonies.

All species when taken together represent a continuum of evolutionary strategies, patterns of development that assure each species a unique niche, a function, as well as a place. On one end of the spectrum are the K-types, named for a component of an equation that describes how populations grow. The Ks are homebodies, species who live and die in one spot and whose lives are affected, controlled, by their own numbers. For example, progeny of western hemlock, a tree of the deep forest, only succeed when a parent dies, leaving a space in the already occupied territory. What a dreary existence this must be, always competing, excluding, while even your children wait hopefully for the day they too can have a patch in the sun.

On the other extreme are the r-types, again named after the same equation. These species are nomads, even gadabouts. Like weeds, they use rich, new places—fertile habitats created by disturbance, a catastrophe for the homebody. They live shorter lives than homebodies and spread their progeny far and wide, quickly kicking them from the nest, showing them the door, casting them to the wind.

Most species are neither total K nor absolute r, but some combination of the two. Searocket, for instance, is a small annual plant in the mustard family. Its petals are yellow and arranged like a cross, as are those of all other members of its family. Because it inhabits the first sand dune on the beach, searocket develops firm, sturdy branches and thickened leaves to brace against the ocean's salt spray. It's moved steadily up and down the Pacific coastline ever since it was first noticed near San Franciso Bay over six decades ago. Searocket produces two types of seed: one that attaches firmly to the mother-plant, and another that falls away easily—an adaptation allowing these seeds to float away on the tide. The seeds attached to the parent remain buried through winter storms with the battered and broken remnants of their mother to germinate in the spring in an already secured safe site. The others drift away to find, as luck and their genes would have it, a place of their own.

I'm the one who moved away from parents, grandparents, uncles, and aunts, from our farm in Washington, to create a new life in California. One of our kids was born in each state as we drifted south along the Pacific coast. More than a decade later, when our parents' shadow seemed less deep, we moved back to Oregon to be closer to them and the structure they still provide. I've worked every day to make this new place safe, somewhere to be, somewhere to stay. So I was pleased when Susan returned home, enrolled in graduate school, and eventually got a job at our local university. But now, unlike her brothers, who settled nearby, she was leaving again. Maybe I'm less a nomad and more a homebody, but it's not my nature to cast kids adrift. Maybe I don't believe in my own genes enough but, like the searocket, I know I can't rely on luck alone.

Rentals were hard to come by in Santa Barbara, a college town. It was a seller's market, controlled by real estate agencies. To me, these agents were like dodder, a spindly orange plant without chlorophyll, its own means of making food. It is a parasite that twists around other plants, twines over them until its host's nutrients are used up and only a dense mat of orange tendrils is left. Susan took them in stride; resiliency, I think, is a characteristic of the young. But maybe like groundsel, she carries her safe site along, adapting her environment as well as adapting to it.

Even though there were few choices, Susan found three places she liked: an older stucco house overlooking the ocean, an apartment in the mission district, and a duplex in the suburbs. She liked the stucco house best, but there was competition and an application process; others wanted the places too. The landlords' agents wanted to sort, select the best renter. The apartment was small but available right away. She would have to wait a few more days to hear about the others. Restless and concerned, she decided on the apartment.

"You'll need a two-thousand-dollars deposit as a cashier's check," said the manager of the rental agency that controlled the apartment, a wiry little man in his late thirties, mustache, red power tie. "Then I'll write up a contract of offer to my client."

"I can't get a cashier's check here yet," Susan replied, "not without a local bank account and I can't get that without a local address."

"Frankly," he said in a cynical tone, "I'm surprised you left Oregon so ill prepared for the way the rental market is here."

Livid at his arrogance, I somehow held my tongue, wondering if the rage showed through my eyes. I hoped it did.

"If I'd brought a cashier's check from my bank in Oregon, who would I have made it out to, and how much would I have made it for?" she asked back.

"Well, how do you do it up there in Oregon, then?" He seemed taken aback by her directness. Perhaps he wasn't used to being talked back to, especially by someone he presumed to be a lowly student.

"We make an offer, write a check, and shake hands on it," I interrupted suddenly, looking straight into his face without blinking.

"Ah-hmmm, well, tell you what, we'll try to help here. We're not really set up for this, but if you'll make a two-hundred-dollar cash deposit, I'll give you a receipt and write up the offer." He printed out a contract, handed Susan a copy, and returned to his desk, shuffling papers while we read.

"Dad," Susan whispered, "what does paragraph two mean?"

"I think it means if you make this offer, you'll be liable for a month's rent even if you choose another place while the owner is thinking it over."

"That's right," the agent replied. He must have been eavesdropping across his desk. "It's a protection clause for my client," he added quickly.

"I need to talk to my Dad outside for a minute," Susan said. "Will you excuse us?" Outside, we talked it over. She wanted to wait to hear about the other two places. She went back in and retrieved her two hundred dollars. If it had been up to me, I'd have turned our whole little caravan back north.

Later that day, Susan located another place, a duplex downtown. But when she called about it, the agent told her that it would be a waste of time to apply. "Technically," the woman said, "you are unemployed and therefore unqualified, because your position with the university here doesn't start for another week." After that, I watched two basketball games on TV that I didn't care about. Lynn and Susan, more productively, interviewed day care centers for Tyler. The next morning, the landlord of the little stucco house called. She'd need to wait another day for the current tenant to move out, but Susan could have the place if she wanted it.

Susan and Tyler now live in California in the stucco house, a sunny place on a mesa overlooking the ocean. Her landlord turned out to be a person, not a parasite. We met him in his home and we shook hands after I wrote a check for Susan's first and last month's rent, my gift to her. Writing it felt like tilling new soil, planting, and knowing your crop is safe in the ground. But the true test of a safe site is whether a seedling—a child—can survive, grow there. Lynn and Susan found a preschool for Tyler quickly, a learning, active place that is run by a church. He slipped into his new niche, adjusted without skipping a beat or pausing a step. As for Susan, we know from her calls that she is making new friends, adapting to her new role at the university. The two of them planted a garden in the courtyard of their new home last week.

Wind Songs

You know you're my lover, when the wind blows I can feel you
—Carlos Santana, *Maria*

I'll be gone for a month, teaching in Argentina. In the last minutes before I leave, Lynn and I look over our just-harvested vineyard from the living room window. The vines have yellowed—the first sign of fall—and it showered a few days ago, settling the dust, relieving temporarily the stress of our normal summers' drought. We talk about small things, the little things around the farm that we both care about—each of us knowing, somehow, their other meaning.

"The rabbits are doing fine since you've moved their hutch." *I'll be okay here while you're gone. You've made sure of that.*

"The dogs are minding the fences better." *Things here are working as they should. You'll be fine. I trust you.*

"The vineyard will be dormant soon." *Winter could be setting in before I get back.* "I'll show you how to damper the stove so the fire will hold all night." *Take care of yourself. I love you.*

I call from Miami and again from Buenos Aires.

The Facultad de Agronomia, where I'm teaching for the next several weeks, is an experiment station for agriculture located in the heart of Buenos Aires. Rows of two-story houses, apartment buildings, traffic noise, and a high white-brick wall surround the half-mile-square campus. The gates stay open during the day, though, so the townspeople who live nearby can cross through. The buildings inside the compound are made from the same whitewashed brick as the wall, but are more weathered. The streets that ramble through campus resemble graveled country lanes more than city boulevards; each is named for a different kind of tree—Calle Magnolia, Calle Sycamore. Doves coo and calandria sing softly in the branches above these unpaved roads. A footpath leads through a vineyard and orchard, alongside open fields of wheat and rye where tractors rumble and students tend their professors' experiments. Honeysuckle twines over every fencerow, inundating the air with its over-sweet fragrance, a contrast to the exhaust fumes and the acrid smell of the city. A group of old men play pinochle each day on a makeshift table of wooden boxes covered with a checkerboard oilcloth in the shade at the corner of Sycamore and Magnolia. Children play soccer across the open fields between the buildings after school, unless goats or sheep are tethered there.

My hosts are Claudio Ghersa and Rolando Leon. Claudio is stout and sturdy, and comes from the cold scrub plains of Patagonia. We worked together in Oregon for nearly three years, until he returned to Argentina two years ago, resuming his teaching job with the university here. His heart belongs to Patagonia though, where he returns each summer to tend the ranch there that he inherited several decades ago. Rolando, a native of Buenos Aires, is agile, adaptable, as much at ease with the tango and opera as he is being among the guanacos and rheas of the not-too-distant pampas. The three of us will travel there tomorrow, into the flooding pampas.

It takes most of the morning to negotiate the city traffic in our rented blue Peugeot van, moving consistently, slowly toward the outskirts of Buenos Aires. Claudio and Rolando have studied the pampas most of their careers. They explain how the low rolling dunes we're passing through form the grassland in this part of Argentina. The low ridges, made from blowing sand, some no more than a meter high, are always crescent shaped and have a companion pond on their lee side. These ephemeral pools fill with the winter storms and can be miles long, but are never very deep. They dry up every summer.

We stop for lunch at Casalins, an abandoned railroad depot with a general store, a gas station, and a restaurant made of dried mud walls and a thatched roof. There are only half a dozen tables inside the restaurant and five are taken. Handmade hunting knives, their handles wrapped in armadillo skin, hang from the restaurant's whitewashed walls. We sit across from a fuzzy-screened television with the American Gladiators on too loud. I stare out the window toward a grass-filled pond where a pair of scissorbills dips, feeding, treading water against the wind. It's stifling in the restaurant. I step outside into the treeless plain, a waving sea of tussock grass. The horizon sways while the muggy spring wind fingers my hair, strokes my graying beard. I lean back against the restaurant's mud wall, captured by the expanse of this place and its motion. One time back home, I climbed a ladder up more than a hundred feet to a platform in the midst of a Douglas-fir forest. The breeze up there was filled with moisture and ozone, like it is here in the pampas. The treetops swayed together in waves like the tussocks do. I held on to the railing of the platform then, to stay balanced, taken by that ocean of swaying treetops.

After lunch, we turn off the pavement on to the raised bank of Canal Nueve, one of a series of unsuccessful attempts to drain the flooding pampas in the late 1800s. The dike road is the only way to an estancia, a fifty-thousand-acre estate where Rolando has an experiment. Along the way, we come up on a tired gaucho riding a forlorn-looking pony across a rusty steel bridge. We turn abruptly on to a narrow jeep track after crossing the bridge, away from the dike. We pass a pair of abandoned ranchos, bunkhouses if we were in the rangelands of eastern Oregon. When the estancia was in its heyday the ranch's foreman and gauchos lived in one of the buildings; the foreman's wife and children lived in the other. A dead sorrel colt lies across the jeep track, next to a barbed-wire gate and the entrance to the estancia. The colt has been dead several days. Blowflies buzz over the carcass; maggots wiggle through it. "They should do something with that," Claudio mutters disgustedly while sidestepping the colt and opening the gate. How, I wonder, can the people here possibly bury every animal that happens to die on their property? What does it matter anyway? In a place as vast, as isolated as this, death probably seems as commonplace as life. Best to leave the carcass for the buzzards and blowflies; let the bones decay back to the soil. Let nature have its way. It's less effort.

The main house of the estancia is built of red adobe and covered by an oxidized tin roof. A buried cistern collects rainwater off its roof for drinking. A stairway leads to a screened-in verandah of white columns and a broad path meanders through stately palm trees, planted among the tussock grasses over a century ago. A single omba tree, perhaps forty feet tall with half a dozen main limbs branching from the ground, sits at the edge of the courtyard. Omba is the only endemic tree of the pampas, but even it was imported by a now-extinct native people who lived in the Chaco to the north centuries ago. Other buildings surround the main house: two mud-brick bunkhouses for the few gauchos who still work the place, a barn for their horses, a brick powerhouse with a diesel generator for electricity, and the butcher's shed that isn't used anymore. Acres of wooden drying racks for cowhides to lie across sit idle in the sun.

The elderly caretaker and his wife greet us on the verandah. While sipping matté, served by their granddaughter—a lovely girl about sixteen years old—we learn that they will move next month. The owners, two brothers who live in Buenos Aires and have never seen the estancia, sold this section of the ranch. The old couple have never lived or worked at any another place, but they will be gone before I return to Oregon.

After matté, we take off our boots beside a rough wood gate at the edge of the courtyard and wade through a shallow slough of ankle-deep, brackish-looking water to Rolando's experiment. The water is warm and it clouds as we walk through the soft mud and cowpies. Mosquitoes swarm shrilly around our ears and above our heads. Rolando describes his study as we wade. How he reconstructed the grassland and the impacts of grazing on it by fencing off twenty-five-acre cattle exclosures like this one over twenty years ago at estancias all across the pampas. Inside the exclosure, the grasses are all native and perennial. They stand upright and grow in tall bunches, tussocks. Outside, the grasses are flat to the ground, growing low, sometimes below the water. I feel humble, awed by this land, recovered from a century of grazing and yet to feel the power of the plow. On the way back to the main house, we disturb a pair of chajas nesting in the tall grass under a fencerow. These gray birds, about the size of an eagle, are so ancient and fearsome that a claw still grows from the elbow of each wing. They live as a couple, perhaps for life. The chajas separate temporarily in flight, but call to each other from a distance—*cha-ja, chaa-jaaa*. It's a lonely cry across a lonely land.

It's dark by the time we return to the dike road on Canal Nueve. There are still fifty miles to go before we reach Dolores and return to the highway. The dike has dry ruts left over from last winter. They make the van rattle as we crawl through the night. It's too dark to see off the side of the dirt road, but we know there's water to our left. What's on the other besides darkness? Our headlights momentarily trap the ghostly images of white-faced cattle grazing along the roadside. We rumble by and Claudio turns off the headlights so the animals won't spook into our path. We drive slowly, alone, through the moonlight night while the ever-present wind buffets the sides of the van.

Rolando begins a lonely, haunting melody in Spanish. His dry hollow voice rises and dips with the wind as it gusts outside the van. The song rolls across the grassland, accompanied only by the wind. It tells of a *carretero*, an oxen driver, who hauled freight a hundred years ago into the pampas using only a cart with two large wooden wheels. He must leave his wife to work alone for months in this desolate land of grass and water. Rolando sings lonesome verse after lonesome verse, until his voice and the rising wind outside the van form a duet that tells the story of a lonely man in a lonely place.

Ay! Paisanita, vuelve a mi amor	Oh little country girl, I'll come back my love
sin ti mi vida no puede estar.	without you my life cannot be.
Las madreselvas se han marchitado	The honeysuckle has wilted
y las calandrias no cantan ya.	and the calandria sing no more.

I awake later that night in Dolores to the rising claps of a spring thunderstorm. The rain pounds hard against my hotel-room window and I remember how it sometimes sheets across the panes of my home in Oregon. I lie quiet listening as the thunder rumbles overhead. I think about the decaying colt returning to the earth in the open rain, about a tired gaucho asleep somewhere in the pampas, and an old man and his wife leaving the only land they know. I remember a single plot of ground restored in this vast land.

As the storm blows through, I remember the chajas calling across the prairie for each other, and reconstruct the ballad that Rolando Leon sang with the wind earlier in the evening. I think of Lynn at home, and wonder how her fire is holding.

I'm back in Argentina after being away for over two years, traveling once again with my friend Claudio. We've stopped at his farm near Conessa for the night. We'll continue on to Baraloche in the morning. It's the shortest day of the year here in Patagonia and the sun will set soon, although it's only four in the afternoon. I peer intently toward a dried-up oxbow of the Rio Negro from this little rise that I'm standing on. It's near the heart of the estancia, the ranch where Claudio grew up, his homeplace.

The wind hits from all sides, pushes itself into my awareness like a cold stare, burns the face, dries the eyes. I pull up the collar of my sheepskin mackinaw and lace it tighter around my neck, tug down on the brim of my felt hat firmly so it sits low across my brow. Only a slight depression now reveals where the ancient campsite had been, a place to rest, hide maybe, escape this endless wind. It's now taken over by willows.

The light's fading. If I stare hard, I can see the solitary light of an oil lantern in the kitchen of the main house, miles away—east, I think. It flickers through the swaying branches of salt cedars planted in the estancia's courtyard three generations ago. Once, twenty families lived on this ranch. It had vineyards, orchards, grain fields, ten thousand steers, and a general store. It's all but deserted now, except for a few hundred head, and a caretaker who lives alone. What's left of the cattle have gone wild, roaming the knee-high creosote bush at will. Used to the free range, they'll die from anxiety if penned up for more than a day or two. Maybe if either of the two brothers had stayed after their father died, the ranch would be different now. Claudio tried for a decade but finally gave up. Even his children can't know how he loves this place. He comes back every summer to work and remember, but it's not enough. Wildness has taken over.

Under my feet near a creosote bush, where the earth's been mounded from centuries of wind-blown silt, are a few shards of flint and a piece of arrowhead. Kneeling to see better in the failing light, I find the ashen flakes of eroded bone on the soil's surface, chips from someone's skull. The wind cuts raw into my sheepskin, pressing deep into the sinew of every joint. I know this wind. I know these sparse hills and gray knee-high shrubs. I could be on my own farm, the homeplace, a continent away. The place I grew up.

I pick up a larger bone barely covered and lying between two bushes, exposed. It's a human vertebra, trampled by decades of cattle hooves, weathered by centuries of rain and relentless wind. If I dig here, I'll find a clay pot or a spear point, something to take back. If I brush away a few more inches of dirt, under any of these gray bushes, there'll be a person—either a man or woman—crouched in fetal position, arms crossed, looking west toward the Andes, their ancient home. They migrated here from the mountains they now face a millennium ago, hunted guanaco on this windy plain, camped in the protection of this oxbow, planted maize and gardens here.

Where do they go, I want to know—these people who love the land, then lose it? Once years ago I dug my hands into the graves of a pioneer family in the Wenas near my childhood home and wondered why the land can't hold on to its people. I asked again the day Dad died, when I climbed our rock butte at the homeplace and took that icy wind in the face. Why must people be buried in it, to keep the land they love?

I fold the bone into my palm along with the flints and arrowhead, hollow out a hole in the ground with my free hand and return the bone and rocks to the earth.

It's dark. I'm lonely. My thoughts turn home.

Backlash

It has been raining most of the night, a solid, hard, ratta-tat-tata kind of rain, not like the mists of Kla-kla-nee or the thunderclaps that form over the homeplace. It startles me into awareness, beating like a snare drum against the blue plastic tarp suspended only inches above my face. I lie awake listening to its repetitions and the rhythm of the foghorn off Slip Point. Green and white lights of other fishermen's boats bob in the darkness, flicker fairy-like in the black sky that merges into the ocean.

Muffled snores emerge from inside the walls of the camper where my brothers-in-law sleep. This is the second day of our annual meat trip, fishing for salmon in the Straits of Juan de Fuca. We come back to this place every year, like the fish we catch. Thankful to be out in the open in spite of the rain, I roll over inside my down-filled bag and sleep again until the aroma of fresh coffee brewing in the camper and the whine of outboards on Callam Bay wake me once more. It's before dawn and the three of us dress quickly, gather up our rain gear, and hurry to the boat with a cup of Ken's caustic black coffee in hand. Ken's coffee always smells better than it tastes. We'll be back with our limits of bright silvers, coho salmon, before noon.

We mooch for a while off Mussolini Rock for chinook—motor off, drifting with the incoming tide, our fishing poles in hand, bait dangling near the ocean's floor. We move in closer, now less than a hundred feet from the squatty shore rock that a fisherman, years ago, named for the World War II dictator. A rip tide usually forms along this section of shoreline, especially during high tide. It brings in herring and other baitfish. But this morning there is no rip, no herring, no chinook, and no luck.

We troll into deeper water and fish for coho using two downriggers, one on each side of the boat. A downrigger has a crank with a thin wire cable rigged on pulleys that are attached along a five-foot pipe. We clip the fishing line to the cable and lower it straight down with an eight-pound lead weight to troll deep and steady. Bob sets one downrigger at forty-eight feet with a green plastic squid, a hoochie, about eighteen inches behind a silver flasher. I lower the other to eighty-six feet. It has a yellow hoochie and gold flasher. When a salmon bites, the line pulls free of the cable. Then you reel in fast to take up the slack line before the fish shakes free. The green hoochie does its job. Soon, we're taking turns passing the fishing pole among the three of us as each fish practically hooks itself. We land at the boat basin with a full limit of silvers just as we figured. It's still early morning.

Back at camp we clean our fish, pack them in ice, and plan the rest of the day. Bob is ready for a nap. Ken wants to shower. "Well ok," Bob cautions, "but don't wash off our good luck." It's our standard joke. I re-rig my salmon pole with a long leader, two treble hooks, and a cheap half-ounce lead weight to fish for greenling off the jetty. I trudge slowly across the piled-up rocks on the jetty, floundering over crusted layers of gull guano and old fish bones until I reach the entrance of the boat basin. The place smells like old fish guts and diesel oil. I snag up on my first cast and lose the whole rig. I rig up again but on the next cast the line hangs up on a rock behind me and backlashes my reel. Goddamn it! I stare into the rat's nest of tangled line.

Annoyed by my misfortune, I figure that if I'm going to unravel a backlash at least I'll go to where it smells better. I lumber back across the full length of jetty. There's a group of fishermen surfcasting off the beach a quarter mile away. I poke along the shoreline across a patch of dry sand, then climb over a pile of driftwood logs to where the first man is fishing. He's dressed in khaki neoprene chest waders and a blue nylon jacket that says Navy Reserve across the front. He isn't catching much so I move farther down the beach and sit on a driftwood snag, still unwilling to tackle the backlash in my reel.

Still farther up the shore is a family. The man is tall, heavy-set, and probably a little over thirty years old. He has on a red flannel shirt with the elbows worn out, bib overalls, and new knee-high black rubber boots. His beard is red, scraggly, unkempt, his complexion ruddy. A boy with a bamboo pole casts into the surf, time after time. He starts each cast by running headlong into the water until it's up to his crotch, then with both hands on the pole he catapults a lure into the incoming waves. It's exhausting to watch him, casting again and again, each time with the same poor result. A fat woman and a girl who's about ten years old tend a fire burning in half a fifty-gallon oil drum that's washed up on the beach. They're roasting a salmon wrapped in a greasy piece of aluminum foil on a makeshift grill of driftwood twigs and discarded metal tubing. The fat woman wears loose-fitting jogging sweats and a torn nylon jacket with its stuffing hanging out. Her stomach nearly touches her thighs as she bends toward the campfire. Her face is flushed. I wonder if she's well. The girl holds a baby wrapped tight in a thin dirty blanket.

It's a meat trip for this family too, but a lot different than my own. I think about the three limits of coho on ice back at camp. There aren't any ice chests here, no camper, no boat, no downriggers, no hoochies. These people aren't here on vacation like me, or even for a day on the beach like the navy guy fishing a few dozen yards away. These people are here for lunch, probably dinner too.

I work at the snarl in my reel, pulling at the tangles with my fingers, tugging on the line to set some of it free. After a while the red-bearded man wades out of the surf, changing his lure for another that he takes from his front shirt pocket. I look toward the woman and girl tending the fire. There's no tacklebox there either. "What's out there?" I ask as he approaches, pointing with my head toward the Straits. I already know that he isn't doing any good.

"Silvers," he says in a voice that matches his course complexion. "You can see 'em in the surf. I caught that one first thing this morning." He points toward the fire.

"Tides out now," I say, fiddling with the reel. I open my pocketknife and slice through the backlash. "Fishing won't be good again until the tide changes."

"Yeah," he says, "I know. We were here last week and did pretty good, caught three, but it was high tide then." He ambles back into the surf flipping the new lure casually into the surf, obviously not expecting much to happen.

Alone again, I think about the three coho I'd just caught this morning. And I think about my sons and our fishing trips to Yaquina Bay for silvers years ago. Those turned out to be meat trips too, but only because we caught so many fish. One year the three of us caught sixty salmon, a full season's limit each, off the sandspit near South Jetty. We used a half-ounce red bolo with a silver flasher. The fish only bit if the flasher was silver, not bronze, gold, or some other color—utterly plain silver. It took a while to figure that out. I wonder if a bolo would work here too?

There were so many fish then; I thought the run would last forever. But it didn't.

"Ain't fishin', huh?" It was Redbeard again, rustling me out of my daydream.

"Naw," I reply. "I've been out today already," nodding toward the boat basin. "I was killing time fishing for rockfish off the jetty, but backlashed my reel. Best not press my luck; the fish cops will be out here for sure. They've been checking punch cards at the dock all morning."

"Dad, this lure ain't no good." It's Redbeard's son, a lanky dark boy about twelve years old with sandy straight hair that hangs over his ears. His jeans are worn through at both knees and he's wet almost to his waist.

"No it ain't," Redbeard growls, irritation showing in his voice. "Tide's wrong, that's all. Nothin' I can do about that, 'cept wait. Go over there by the fire and eat if you're tired." The boy hesitated. "Damn it, go eat!"

I rummage through my tacklebox. "Here, try this," I say, holding out a red bolo with a silver flasher toward the boy. "These used to work pretty well in Yaquina Bay for silvers." The boy reaches for the lure, then stops and looks apprehensively at his father. Redbeard nods slightly and the boy snatches the lure from my hand and runs into the surf.

"From Oregon, huh?" Redbeard says. "I used to live in Oregon, out by Clatskanie. Had a good job down there, settin' chokers around logs for Longview Fiber 'till the goddamned enviros shut us down! Shut us down, that's for goddamn sure."

I look into my now half-full spool of line, fiddling with the reel, wondering how to shift this conversation from where it's going. I want to reshape this conversation, like one of my grapevines, into something more productive. I'd like to say he's got it wrong, that it wasn't environmentalists who shut down the cut. It was the timber companies and Forest Service that did the over-cutting; the enviros only called attention to it. They managed to stop the cut while there were still a few old-growth tree stands left. I wanted to tell him that the old forest is gone, depleted, run out of trees. That logging had to slow down eventually for the regenerating forest to catch up with the cut. But I don't. I know it's pointless to argue, and Redbeard wades back into the surf.

I feel sorry for Redbeard and his family. Setting chokers is the least skilled, worst paid, and most dangerous job in logging. But I don't know why he got laid off from Longview. Maybe because of poor work habits, or because someone else had more skill, more knowledge, more ambition. Maybe the company laid off all their choker setters. Maybe he's a company castaway, an expendable, a pawn in the high-stakes game for wood and money left now with few skills to adapt to a new way of life. But I don't know.

"Dad, Dad, come quick. I've got one." It's the sandy-haired boy yelling. He stumbles backward out of the surf with a twenty-inch silver on his line, forgetting to reel in the excitement. The fish flops helplessly on the beach while Redbeard unhooks the salmon and whacks it across the skull with the handle of his hunting knife. The boy runs up the bank to his mother at the campfire, holding the middle of the fish with both hands.

Later, Redbeard walks over to my snag. "Thanks," he grumbles. "That was Sean's first fish, ever," and he holds out his hand. I shake it, surprised by his gesture.

On the way back to camp, thinking about lunch and a shower, I hear someone running up behind me. It's Sean. He stops as I turn and silently holds out the red bolo, his eyes down. I'm pleased that he's caught up to me, but wonder if his father sent him to return the lure or if he came on his own.

I wish that I could sit with this boy for a while, like I've done with my sons and daughter. I'd like to point toward the ocean and tell him that he has options, perhaps as many as silvers in this bay. If he continues to work, he can learn to catch them. He started today using my bolo lure. That even though his life now is being shaped by the decisions of others, he needn't be limited by the life his father chooses. He can go to school every day, learn, study hard, or not. His life can be like a well-pruned tree, but the form it takes ultimately depends on him, the opportunities he takes and the ones he decides to let go.

Instead I say, "You keep the lure, Sean. There's lots of fish in this bay; keep learning to catch them." He looks at me quizzically for an instant, then turns and runs down the beach to his family.

I walk back to camp with the backlash still knotted in my head.

Planting Again

Almost a century ago the U.S. Bureau of Reclamation diverted half of the Tieton River through three mile-long tunnels and about ten miles of cement flume into the North Fork of Cowiche Creek. Irrigation water still collects behind the dam at Rimrock Lake, a reservoir in the Cascades, then it's channeled into the first and highest tunnel; it busts out of the mountain a mile later into a ten-foot-high cement flume. The flume winds around the low hills for miles, then gushes through the middle tunnel—a second ridge—and into another flume on the other side. The water again wanders across the foothills for miles until it tumbles through the final mile, the lowest tunnel, and splashes icy cold, spitting, into the North Fork above Adrian French's farmhouse. The irrigation project took a decade from start to finish using draft horses, immigrant workers, and lots of dynamite. It cost half a million dollars back then and caused at least one lynching, apparently of a contractor who developed a permanent cash-flow problem.

Until a decade ago water from the tunnels dallied in French's Canyon behind a low diversion dam that split it into three broad canals, each named for one of the first three letters of the alphabet. The canals, ditches, meandered over the dry land, gracefully crossing through the sagebrush and prickly pear of the high desert, twisting around steep bluffs of black basalt boulders. Orchards and hayfields replaced the desert sage.

Grandpa bought our farm three years before the Tieton River Diversion Project was finished from Tom Donnelly, the man who homesteaded it. Grandpa divided the forty acres into five irregularly shaped blocks, based on the topography of the land, to irrigate the apple orchards he intended to plant. The irrigation project delivered water from the A-ditch through an open three-foot-deep trench that ended in a cement box that Grandpa built on the highest corner of the farm. From there, it flowed underground through four six-inch asbestos and tar pipes, mainlines, one for each block of apples. He walked over the whole farm twice a day, setting water into shallow ditches that ran the full length of every tree row.

During winter and early spring,, the water is shut off above the highest tunnel so Rimrock Lake can fill as the snowpack melts in the Cascades. One early March day when I was a boy, three friends and I rode our bikes up the tunnels to where they begin. We met at Newland's drugstore and rode the five miles up French's Canyon to the lowest end of the tunnels. We hesitated at its gaping black mouth, peering into the long tube, into the middle of a mountain. The tunnel is five or six feet around but the flashlights that we'd taped on to our handlebars failed to pierce its secrets for more than a few feet. I remembered foreboding stories that we'd all been told about these tunnels—of bear and bull elk trapped inside with nothing to eat, frenzied by starvation, and of hermits—ferocious ragged little trolls who lived in the tunnels all winter long. We'd heard tales too of open spillways that plunged hundreds of feet through caverns in the mountains to the river below.

Bats, frightened by our flashlights and bike noise, fluttered just above our hair to escape into the sun as we passed through the tunnel's waiting open jaws. Inside, we pedaled uphill in single file, never speaking, hunched over our handlebars, eyes fixed on the frail lights we'd strapped to our bikes. We rode intently toward a speck of light that appeared magically at the far end of the tube; watched it grow to the size of a dime, nickel, quarter, silver dollar. Then we burst out of the mountain into the flume on the other side, squinting from the harsh winter sunlight.

At lunchtime we stopped at the bottom of the highest tunnel, the last one on our uphill trek. We stood on our bike seats and pulled ourselves over the top of the flume on to the cliff's edge and ate our sandwiches under a scraggy-topped ponderosa. Ant-sized cars crawled toward the lake on the river road below. We lowered ourselves back into the flume after lunch, pedaled a quick mile through the last tunnel, and came out on a rocky creekbed surrounded on both sides by steep snowy slopes of fir, pines, and larch. We left the bikes in the dry creek and hiked another mile to where a fish screen across the stream blocked our path. The lake was only a few hundred yards farther ahead. We could see the dam, but with the expedition complete we ran back to our bikes, then raced hooting and hollering through the now familiar tunnels and cement passages until we were home.

I swam in the A-ditch when I was a kid and hunted trout in its puddles with a pitchfork after the Project turned off the water every winter. Now the old canals are filled in, buried, replaced by a hollow underground network of five-foot-diameter steel pipes—conduit that originates behind a much higher diversion dam that floods most of French's Canyon. This lake without a name, now permanently filled and fenced with chain-link and razor wire, cuts off every access to the tunnels even during winter when no water flows through them. Farm ponds, cisterns, and wells dried up without seepage from the canals to keep them full. The North Fork is, once again, an ephemeral desert stream, fed only by a few hidden springs under the cottonwoods and willows along its banks. The irrigation water is measured out now too, metered carefully by the Project. The homeplace gets exactly 24.8 gallons of water per minute from April 1 through September 30. It arrives underground at the same cement box where Grandpa started his mainlines, but now it comes under two hundred pounds of pressure.

Grandpa stood barely five and a half feet tall. His shock of silver hair and full white mustache contrasted sharply with skin darkened and weathered by years of sun and hard orchard work. He usually wore a faded red plaid shirt, blue-striped overalls, and a battered gray felt hat. He never seemed young to me, shuffling on worn-out feet over the tractor roads that separated his four blocks of orchard, hands held tight behind his back. He smoked a curved ivory pipe and smiled with his eyes. I wonder what Grandpa would think now of the isolated tunnels, nameless lake, and buried canals. Would he mourn the passing of the old A-ditch, miss the wells and ponds that he and Dad dug and depended on? Or would he like the change, marvel at the efficiency of the Project's new water system, appreciate its pressure, its progress? Does it matter?

Dad sold most of our water rights to keep the farm going when times got bad, believing he could keep on irrigating the orchards from our pond and wells forever, even without a full allocation from the Project. He was right, except for the Project engineer's plan to pipe the water system, to be more efficient, less wasteful. It's tough to farm in Tieton without enough water. The desert tries to reclaim its land.

All summer and fall after we first took over the homeplace, my sisters, brother, and I worked one three-day weekend every month to replace the old mainlines. The worn-out asbestos and tar pipes that Grandpa had laid were rotting away from age and decay. The Project's new pressurized system would have blown the old pipes out of the ground even if they could have been fixed one more time. Our new mainlines are made of two-inch PVC, and divide the farm evenly into seven three-acre blocks, each irrigated by its own solenoid valve. We use black polyethylene hoses and precision water emitters that drip out a gallon of water per hour to all our trees. No one walks the farm anymore to set the water every day like Grandpa and Dad did; a computer does that now.

It bothers me though, this lack of human involvement on the land, this abolition of farmer from the farm. I feel separated, cut off, inaccessible to the land like the tunnels are now to the people who depend on them. I hope I didn't give up too soon on Grandpa's legacy, on the farm's subtle curves and contours. Then again, each of our generations changed the farm, making improvements and mistakes, adapting the land as it, and the times, shaped each of us. First there was only sagebrush and cheatgrass, then canals and an orchard appeared. Next came wells and a pond, and now polyethylene tubes and a computer.

It felt good to plant trees back on the homeplace the summer after Dad died, to put life back into the ground. Carol, my sister, and I got an early start on that first day of planting, just before sunrise. She's tall, quiet, and has dark hair and skin like me. She lives in Spokane and loves the farm, the way it was. We drank hot coffee from her silver thermos that morning, both of us double wrapped in a sweatshirt and work jacket, waiting for the reluctant sun to peek over our rock bluff. It's serene on the farm in the early morning. The wind always lies still, the cool silence broken only by the distant calls of sentinel quail collecting their flocks across the bluff. We planted fifty seedlings by shovel before the others joined us. We could have used the two-person power auger that I'd brought along, but decided not to. Neither of us wanted to break the spell that the early morning solitude cast across the farm, and us.

But Polly, our youngest sister, would have nothing to do with our low-tech approach for planting eight hundred and fifty more trees. So by mid-morning, she and brother Joe powered up the auger while Carol, sister Annie, sister-in-law Marion, and I planted the seedling trees. We'd grown apart over the years; each of us has our own families now, separate jobs, vocations, interests. Carol and I live farthest away from the homeplace, and from each other. Joe and Marion live down the valley, seventy miles away. He'd been a pretty good running back in high school, but I never saw him play. I was in California then. Polly was born the year I left home for WSU.

I germinated the seedling trees from nuts that Carol collected in Spokane, and we planted the little trees the same way Dad had taught me three decades earlier when sister Babi and I helped him put in his cherry orchard on this same piece of ground. We pushed each walnut seedling into a freshly augured hole, pulling in dirt from around its edges, and tamping it down with a hunk of two-by-four. Then we pulled up lightly on the seedling's stem to be sure the roots grew down, shoveled in more dirt, and stomped on it to make a small basin. I poured a can of water into each little basin, sealing the tree roots from the air. Air kills young roots.

Later, with the others still planting, I rolled out the mile and a half of black polyethylene driplines that it would take to irrigate our nine hundred trees. Each dripline snaked along its tree row and finally connected into the mainline that we'd buried the year before. My work went slowly; the tubing was rolled so tight that it coiled and twisted like soft spaghetti when I cut its strapping. The sun burned down. It penetrated deep into my straw hat, scorched through my T-shirt. There wasn't enough drinking water. I looked for our cool Tieton breeze and wondered why it deserted us. We ended the day dirty, tired, and thirsty, questioning the wisdom of what we were trying to do.

The next day, Carol and I got an early start again and planted the last hundred trees by shovel before the others arrived. Later, with the trees planted, we all poked, pushed, or prodded the nickel-sized plastic water emitters into the plastic driplines, one on each side of every tree—eighteen hundred in all. The day's heat caught us again before all the emitters were in. Sweating and feeling dejected, we began to realize that it would be late in the day before I could turn on the water, if then.

It was near evening when I finally knelt at the junction box where the Project's pipe hooks to our mainlines. Slowly, deliberately, I turned the control valve to release the first water on the farm in a decade. I heard it surge past my hand, felt it vibrate against each closed valve as it plunged toward the end of our pipes. I felt the pulse of the farm as our water rushed again through its veins, sensed its heartbeat. From across the field, I saw Joe kneel to open each valve that sent the cool liquid into the driplines and to the trees we had just planted. I held my breath in anticipation, expectation. It won't work, I thought; lines will blow, valves will stick, emitters plug.

The emitters hissed as air cleared the tubes, then sputtered. Water trickled slowly from them in unison. Then, unexpectedly, a splice in a dripline blew out and Joe, who was closest, ran to fix it. But instead of crimping off the line, he lifted the hose over his head so the water sprayed across his shoulders and across his back. Annie joined him, then the rest of us and we danced together under the umbrella of our own rain—holding each other's hands overhead, fingers interlocked, surrounded by our trees and the wonderful smell of water seeping into this ground that had been too long dry.

As we heal ourselves, we heal the land, I think later that evening back at the old house. Or is it the other way around?